ASHEVILLE-BUNCOMBE TECHNICAL INSTITUTE

NORTH CAROLINA
STATE BOARD OF EDUCATION
DEPT. OF COMMUNITY COLLEGES
LIBRARIES

D1457329

DISCARDED

NOV 2 0 2024

The Radiation Chemistry of Water

This is Volume 26 of
PHYSICAL CHEMISTRY
A series of monographs
Edited by ERNEST M. LOEBL, *Polytechnic Institute of Brooklyn*

A complete list of the books in this series appears at the end of the volume.

THE RADIATION CHEMISTRY OF WATER

IVAN G. DRAGANIĆ

AND

ZORICA D. DRAGANIĆ

Boris Kidrič Institute of Nuclear Sciences
Vinča, Yugoslavia

ACADEMIC PRESS   New York and London   1971

Copyright © 1971, by Academic Press, Inc.
ALL RIGHTS RESERVED
NO PART OF THIS BOOK MAY BE REPRODUCED IN ANY FORM,
BY PHOTOSTAT, MICROFILM, RETRIEVAL SYSTEM, OR ANY
OTHER MEANS, WITHOUT WRITTEN PERMISSION FROM
THE PUBLISHERS.

ACADEMIC PRESS, INC.
111 Fifth Avenue, New York, New York 10003

United Kingdom Edition published by
ACADEMIC PRESS, INC. (LONDON) LTD.
Berkeley Square House, London W1X 6BA

LIBRARY OF CONGRESS CATALOG CARD NUMBER: 78-154396

PRINTED IN THE UNITED STATES OF AMERICA

Contents

Preface ix

Acknowledgments xi

CHAPTER ONE. **Historical Survey of the Radiation Chemistry of Water**

I.	Present Status	1
II.	Literature	2
III.	Early Period: Changes Are Observed in Solutions in Which the Radiations from Radioactive Substances Are Absorbed	3
IV.	"Photochemical" Reactions Induced by X Rays	5
V.	Free Radicals as Carriers of Changes in Irradiated Water	9
VI.	Pulse Radiolysis	14
	References	15

CHAPTER TWO. **Interaction of Ionizing Radiation with Water and the Origin of Short-Lived Species That Cause Chemical Changes in Irradiated Water**

I.	Absorption of Radiation in Water	23
II.	Excited and Ionized Water Molecules	29
III.	Fate of Electrons Produced in the Ionization of Water	35
IV.	Spatial Distribution of Active Species	36
V.	Brief Survey of Important Events Leading to the Formation of Primary Free-Radical Products in Water Radiolysis	38
	References	42

CHAPTER THREE. **Primary Products of Water Radiolysis: Short-Lived Reducing Species—the Hydrated Electron, the Hydrogen Atom, and Molecular Hydrogen**

I.	Properties of the Hydrated Electron	47
II.	The Kinetic Salt Effect as Evidence for the Negative Charge on the Hydrated Electron	57
III.	Evidence for Two Kinds of Reducing Species from Competition-Kinetic Experiments	60
IV.	Atomic Hydrogen	63
V.	Primary Molecular Hydrogen	76
	References	81

CHAPTER FOUR. **Primary Products of Water Radiolysis: Oxidizing Species—the Hydroxyl Radical and Hydrogen Peroxide**

I.	Properties of the Hydroxyl Radical	91
II.	Forms of the Hydroxyl Radical in Irradiated Water at Various pH's	101
III.	Relative Rate Constants of Hydroxyl Radical Reactions; Competition Kinetics	103
IV.	The Hydroperoxyl Radical	108
V.	Primary Hydrogen Peroxide	111
	References	116

CHAPTER FIVE. **Radiation-Chemical Yields of the Primary Products of Water Radiolysis and Their Dependence on Various Factors**

I.	Remarks Concerning the Radiation-Chemical Yields of Primary Species	123
II.	Effect of Scavenger Reactivity	127
III.	G_R and G_M Values for γ-Irradiated Water at Neutral pH	130
IV.	Effect of pH	140
V.	Primary Yields in D_2O	144
VI.	Effect of Linear Energy Transfer	150
VII.	Effect of Dose Rate	155
VIII.	Effect of Temperature	157
IX.	Effect of Pressure	159
	References	162

CONTENTS

Chapter Six. Diffusion-Kinetic Model

I.	Basic Assumptions	171
II.	Some General Cases of the Diffusion-Kinetic Model	172
III.	Theoretical Predictions and Experimental Observations	177
	References	188

Chapter Seven. Radiation Sources and Irradiation Techniques

I.	Radiation Units with ^{60}Co	191
II.	Pulsed Electron Beams	194
III.	Positively Charged Particles	197
IV.	Some Other Irradiation Techniques	200
V.	Preparation of Samples for Irradiation	203
	References	206

Chapter Eight. Aqueous Chemical Dosimeters

I.	Chemical Change as a Measure of Absorbed Dose	211
II.	Ferrous Sulfate Dosimeter (The Fricke Dosimeter)	216
III.	Ceric Sulfate Dosimeter	219
IV.	Oxalic Acid Dosimeter	220
V.	Some Other Systems Used in the Radiation Dosimetry of Water and Aqueous Solutions	222
	References	224
	Author Index	227
	Subject Index	235

Preface

Since the appearance of the first monograph on the radiation chemistry of water and aqueous solutions, written by A. O. Allen a decade ago, significant progress has been made in many areas of this branch of chemistry. A large number of radiation-chemical reactions have been studied, new reactive species discovered, and, at present, water radiolysis should probably be ranked among the best-understood domains of radiation chemistry.

The radiation chemistry of water and aqueous solutions is a comprehensive subject, on which about 3000 scientific, technical, and review articles have been published. The present book deals with radiation-induced changes in water and, to the extent necessary to explain the behavior of irradiated water, with changes in aqueous solutions. An up-to-date book on radiation chemistry that will systematically review numerous radiation-chemical changes observed in different aqueous solutions still remains to be written.

Various problems encountered in physical chemistry, chemistry of radioactive solutions, radiobiology, nuclear technology, or applications of radiation in industry call for a better understanding of the radiation-chemical behavior of water. The present book is intended for such a reader, who is also presumed to have some knowledge of chemistry and of the properties of high-energy radiation.

As can readily be seen from the contents, we are primarily dealing with short-lived species, the hydroxyl radical, the hydrated electron, and the hydrogen atom, which cause the chemical changes in irradiated water and aqueous solutions. We have considered their origin (Chapter Two), their properties (Chapters Three and Four), and the dependence of their yields on various factors (Chapter Five). The diffusion-kinetic model of water radiolysis is treated separately (Chapter Six). Kinetic analysis to check proposed reaction mechanisms is certainly one of the most effective approaches in water-radiolysis studies. This is why some general cases (in Chapters Three through

Five) are considered in more detail than would be justified by the simple kinetics alone. Unlike the customary practice, radiation sources (Chapter Seven) and dosimetry (Chapter Eight) are considered in the last two chapters. The reason for this lies in our desire to give some details which would apparently have overloaded the introductory part.

Acknowledgments

The authors are indebted to the Boris Kidrič Institute of Nuclear Sciences (Vinča, Yugoslavia) for facilities, in particular for the possibility of preparing the manuscript also within the framework of their regular duties at the institute. Our thanks are also due to the Centre d'Etudes Nucléaires de Saclay (France) where, during a wonderful sabbatical leave, the final version of the manuscript was written.

We owe a great deal to Drs. J. Sutton (C.E.N., Saclay, France), A. O. Allen (B.N.L., Upton, New York), J. Bednař (Řež, Czechoslovakia), and V. Marković (Vinča, Yugoslavia), who helped us by criticism of our manuscript. We also wish to express our deep gratitude to our colleagues from the Radiation Chemistry Laboratory of the Institute in Vinča, who collaborated with us in various ways in preparing the manuscript.

It is a pleasure to acknowledge the participation of Mrs. S. Subotić in the translation of the manuscript into English.

This book could well be dedicated to Sonia and Milan, for they had so often to wait patiently for their parents.

Radiation chemistry is a branch of chemistry primarily concerned with chemical effects of high-energy radiations such as those made available by radioactive substances, high-energy machines, and nuclear reactors. The first radiation-chemical change was observed as early as 1895 by Roentgen, when he established the existence of a penetrating, invisible radiation—X rays—able to fog a photographic plate, that is, able to produce a chemical effect. The term "radiation chemistry" was proposed for the needs of the Manhattan Project, the secret atomic energy research program carried on in the United States during World War II. In this connection Professor Milton Burton, now at Notre Dame University (Indiana), has written:

> In May 1942, the title radiation chemistry did not exist ... the name photochemistry proved awkward; there was too much confusion, too much overlap of interest, and I sought an appropriate name for an area that we quickly realized had existed for 47 years without any name at all. The name radiation chemistry came out of the hopper; I didn't like it; I asked Robert Mulliken's advice. He couldn't think of anything better and, with that negative endorsement, the old field received its present name [*1*, p. 87].

CHAPTER ONE

Historical Survey of the Radiation Chemistry of Water

I. Present Status

Radiation chemistry is a comprehensive subject embracing many areas of research. This book is concerned with the radiation chemistry of water, a field dealing with the chemical changes induced in water by the absorption of high-energy radiation. At present, the main efforts of investigators are directed toward a better understanding of the origin of reactive species responsible for the chemical changes observed in water and aqueous solutions. Of special interest are the nature of free radicals created during irradiation and rate constants for reactions of these species with one another or with ions and molecules present in the solution. Interest in the action of radiation on living matter, the use of water and aqueous solutions in nuclear technology, as well as the possibility of using high radiation doses in industry, have also initiated various specific studies in this field of radiation chemistry. In addition to the mechanisms of radiation-induced chemical processes, the practical aspects are treated: problems of chemical dosimetry, design of radiation sources and suitable arrangements for irradiation, as well as methods devised for the observation of reactions during or after irradiation.

As is often the case in other disciplines, the first stages of radiation chemistry were restricted to a description of observed phenomena and their interpretation in terms drawn from related branches of chemistry. At the present stage, however, the results of investigations in this field have already begun to make considerable contributions to other branches of chemistry. This is especially true with regard to chemical kinetics. The establishment of mechanisms in chemical kinetics is always difficult and often problematical. Radiation chemistry is not an exception in this respect, but it has a certain advantage over investigation of the kinetics of chemical reactions induced in other ways. This has become especially evident during the past few years, since it is now possible to introduce into the system studied energy "packets" of penetrating radiation, which are sufficient to produce measurable amounts of reactive species in one billionth of a second. But it is not only the contribution to direct observation of very fast reactions that is important. At present, the development of stationary-state kinetics, as well as the kinetics of competition processes, probably is due more to studies in the field of the radiation chemistry of water and aqueous solutions than to those in any other field.

II. Literature

Figure 1.1 shows the number of publications on the radiation chemistry of water and aqueous solutions from the first reports to the end of 1969. These data, taken from two bibliographies [2,3], show that the total production of articles, monographs, and textbooks does not exceed 3000 titles. In compiling this list, account has also been taken of some papers of which the titles do not directly or exclusively bear on the field in question, but the contents of which are, at least in part, of interest for the field of the radiation chemistry of water. In what follows, individual publications are referred to in the usual way. On the average, only every fifth reference is cited in the present book. The reason for this lies in our desire to give an account of the basic aspects and main problems concerning liquid water rather than a review of all that has been done in this field, especially with aqueous solutions. We refer the reader who is interested in a more detailed history of the radiation chemistry of water and aqueous solutions to the excellent papers of Hart [4] and Allen [5-7].

Illuminating accounts of the radiation chemistry of water and aqueous solutions have been given by Allen [8] and Hart and Platzman [9]. These books are still precious sources of information. The series of monographs "Action chimiques et biologiques des radiations," edited by Haissinsky [10], contains many contributions that are extremely useful for all those interested in the subject treated in the present book. This is also the case with the more recent series "Current Topics in Radiation Research," edited by Ebert and Howard [11], and "Advances in Radiation Chemistry," edited by Burton and Magee [12]. Different books on nuclear chemistry [13,14], radiation chemistry [15-17], and radiation biology [18] also give very useful texts in their chapters on water and aqueous solutions. Pulse radiolysis is considered in great detail in two monographs [19,20]. Shubin and Kabakchi [21] have made an original

III. EARLY PERIOD

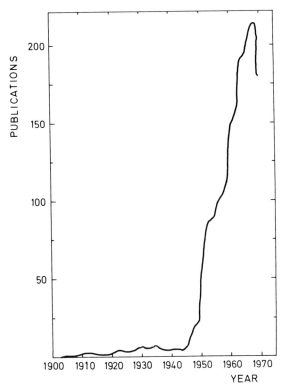

Fig. 1.1. The number of annual publications on the radiation chemistry of water and aqueous solutions.

approach to different theoretical aspects of the radiation chemistry of water. Several symposia on radiation chemistry, containing valuable contributions, have also appeared in book form [22–32].

III. Early Period: Changes Are Observed in Solutions in Which the Radiations from Radioactive Substances Are Absorbed

The earliest days of the radiation chemistry of water were also those of radiochemistry. The discovery of radium was reported by the end of 1898; and less than 5 years later, Giesel [33] quoted the observations of Runge and Bodländer that a gas consisting of hydrogen and oxygen is constantly liberated from aqueous solutions of radium bromide, the solution becoming brown because of the generation of bromine. It seems, however, that Ramsay [34] was the first to carry out systematically a series of complete experiments in connection with "the chemical action of radium emanation." In the title of his article, published in 1907, we find "the action on distilled water." This work and the papers of Cameron and Ramsay [35,36], which followed it in the course of 1908, contain a number of quantitative data among which we should point out those on the generation of H_2 and O_2 under different conditions. The next year Debierne [37] reported the results of his measurements on water decomposition and drew attention to the cause of this phenomenon. He pointed out the different possibilities: The decomposition may be caused directly by radium emanation, by the decay products, or by the action of the radiation itself. His conclusion was that the decomposition results from the action of the penetrating β and γ rays, which was proved by a simple experiment. The author used as the radiation source a sealed ampul of radium chloride. He placed this in a vessel containing distilled water which was connected to a manometer. The system was not in contact with air, although the water was not degassed. The increase in pressure, measured by the manometer, was regular and proportional to time. Kernbaum [38] made quantitative measurements of the gases generated, with the object of checking whether their ratio is the same as in the electrolysis of water or, as shown by Ramsay, the molecular hydrogen is produced in excess. He found the latter to be true, but established as well the presence of hydrogen peroxide. Quantitative measurement led the author [39] to the assumption that water decomposition proceeds in accord with the equation $2H_2O = H_2O_2 + H_2$.

In 1913, Duane and Scheuer [40] presented, in a complete study, their data on water decomposition under the action of α rays.

By present standards these radiation sources, seldom stronger than 100 mg radium, were very weak. Nevertheless, in the literature of that time we also find observations of other changes caused by radiation in aqueous solutions. In this connection the numerous experiments of Kailan [41–46], performed at the Institute for Radium in Vienna, should be pointed out. In these articles we find, among other things, data on reduction of trivalent iron, decomposition of oxalic acid, and generation of halogens in solutions of alkali halides.

Low radiation intensities, as well as the relatively insensitive methods of chemical analysis available at the time, reduced these investigations mainly to qualitative observations of the chemical changes or, less often, to semi-quantitative measurements of gross effects. It may be said that almost no efforts were made to explain the origin of the observed changes. An exception, certainly ahead of the time, was Debierne's [47] hypothesis of the formation of ionized water molecules during the passage of radiation: The reactions of these ionized molecules lead to the formation of H and OH species. Their reactions result in water reformation and the formation of molecular products, H_2 and H_2O_2. Although it appears from the present state of affairs that many of the reactions and species introduced by Debierne do not seem likely, his hypothesis contains the main elements of present theories: ionization of H_2O molecules, formation of reducing and oxidizing free radicals, and recombination of these last to give H_2 and H_2O_2, respectively.

IV. "Photochemical" Reactions Induced by X Rays

In the years following World War I, photochemistry was in vogue. Therefore it is not surprising that chemists arrived at the idea that, in addition to ultraviolet (uv) radiation, high-energy photons provided by the Roentgen apparatus might also be used to induce chemical reactions, especially since at that time quite powerful X ray sources began to be constructed. As may be seen from some of the papers published at that time, photochemistry was considered to have extended its area of research to photochemical reactions induced by X rays. The studies of Fricke and his collaborators, reported in the 1930s, were the first to call attention to the essential difference between processes induced in water and aqueous solutions by the action of X rays and those induced by low-energy photons. In the case of uv radiation, the direct action of a photon on a molecule is in question. But in the case of X rays the effect is indirect: X Rays "activate" water molecules, and the latter lead to the observed chemical changes through their reactions with the ions or molecules of the substances present.

The intense X-ray sources of the 1930s represented in fact a veritable revolution in irradiation technique. They placed at the investigators' disposal radiation sources equivalent to thousands of curies of radium. Moreover, the simplicity of irradiation and the absence of inconveniences such as those encountered in work with natural radiation sources were of no less importance than the possibility of quickly attaining radiation doses sufficiently high for chemical changes to be reliably determined. Nevertheless, the first impetus to the use of these possibilities for the study of chemical changes in water and aqueous solutions did not come from chemistry, but was closely associated with the study of biological changes. Biologists knew that the activity of a protein decreases under irradiation of its aqueous solution. Fricke and Petersen [48] showed that hemoglobin—a protein containing iron in the form of ferrous ion—is transformed into methemoglobin on irradiation in aqueous solution. Methemoglobin contains iron in the ferric form, and the considerable effectiveness of the oxidation reaction seemed curious, especially as there are only four ferrous ions in a molecule of the protein the molecular weight of which amounts to 68,000. The results indicated that it is unlikely that a direct action of radiation is in question, but that the change proceeds indirectly, via the primary action on water. This assumption was also favored by the fact that the amount of the methemoglobin produced is almost independent of the hemoglobin concentration in the solution irradiated. This observation was soon followed by that of the oxidation of ferrous ions in a simpler system—solutions of ferrous sulfate in 0.8 N H_2SO_4. The conclusion drawn by Fricke and Morse [49,50] was unambiguous: The oxidation of ferrous ions is a secondary effect. It is a consequence of the action of active water molecules formed by interaction with secondary electrons produced by X rays. Calibration of the radiation intensity was carried out by means of ionization chambers filled with air. Therefore, the sulfuric acid concentration (0.8 N) was adjusted in such a way that the solution irradiated should have the same absorption characteristics for X rays as air. In these articles, curves are presented which show the effect of the dose absorbed and of the initial concentration of ferrous ions on the amount of ferric ion produced. This solution is exactly the same as that used at the present time in chemical dosimetry and known as the Fricke dosimeter (see Chapter Eight, Section II).

There was no doubt that radiations from radioactive substances decompose water and that chemical changes occur in water under their action or, as shown by Fricke, under the action of X rays. Therefore, the statement of Risse [51] that he could not prove the existence of H_2 and O_2 in a carefully purified and degassed water subjected to X rays was rather puzzling. The inability of X rays to decompose water was also confirmed by Fricke and Brownscombe [52], using more sensitive and precise methods of analysis. There was no explanation of this fact, and it was only 15 years later that Allen was able to provide one (see Section V). Meanwhile, a number of further observations were carried out on irradiated water. Fricke [53] established the existence of measurable quantities of hydrogen peroxide, dependent on acidity of the medium, when oxygen was present in irradiated water. Fricke and Hart [54] showed that pH also affects the yield of ferrous ion oxidation: These decrease with increasing pH no matter whether the oxidation is carried out in the presence or absence of oxygen. In this last case the total number of ions oxidized by a given dose is less than that in the solution where oxygen is present.

Before proceeding to a concise review of other results obtained in this period, which practically lasted up to the first years of World War II, several important facts should be stressed to which attention was called by the previously mentioned experiments on oxidation of ferrous salts: The yield of oxidation is independent, over a wide range, of concentration of the ferrous ions present in the irradiated sample. Consequently, direct action of radiation on water, and through it on the solute, is in question. The number of ions oxidized depends on whether oxygen is present or absent, on impurities, as well as on the concentration of H_3O^+ ions in the irradiated solution. These

IV. "PHOTOCHEMICAL" REACTIONS INDUCED BY X RAYS

conditions are to be taken into account in designing and interpreting any radiation-chemical experiment.

Clark and Pickett [55] subjected to the action of X rays a number of systems which were known to change under the action of uv radiation. Their general observation was that even if changes do occur under the action of X rays, they are nevertheless very small in comparison with the energy absorbed. The authors considered oxidation reactions, which they ascribed to the hydrogen peroxide formed during irradiation, to be the most effective. Among other things, the results of irradiation showed the oxidation of potassium iodide and the reduction of potassium nitrate, confirmed the oxidation of ferrous ions, and yielded information on the behavior of some colloids. Roseveare [56] subjected to X rays the system mercuric chloride–potassium oxalate, also known from photochemical studies. He established the decomposition of oxalate ion and gave data concerning its dependence on various factors. Clark and Coe [57] studied reductions in aqueous solutions under the action of X rays. They showed, among other things, that in ceric sulfate solutions subjected to radiation, tetravalent cerium is reduced. Contrary to what was to be expected, the irradiation of the system uranyl sulfate–oxalic acid, often used at that time as the actinometer for measurement of absorbed energy of uv radiation, did not lead to any visible chemical change. From this the authors drew the conclusion which may serve as a good illustration of the concept of X-ray "photochemistry": "No reaction could be detected, so again it was found that behavior with ultraviolet was not safe evidence for prediction of the action of X rays [57, p. 101]."

Fricke and Brownscombe [58] have presented data on the influence of various factors on chromate ion reduction. The amount of chromate reduced depends on the concentration of H_3O^+ in the irradiated solution, except in media of high acidity. As in the case of oxidation of ferrous ions, the effect observed is independent, over a wide range, of the concentration of chromate present. The temperature of the irradiated solution varied from 5 to 90°C, but no effect on the reduction yield was observed. The effect of the presence of other substances in solutions was particularly studied. The results call special attention to organic substances which, even if they are present in only trace amounts, considerably increase the observed effect.

Fricke *et al.* [59] have studied in great detail the changes induced by X rays in a number of aqueous solutions of organic substances, acids, aldehydes, ketones, and alcohols. They have also given data on the behavior of carbon monoxide in irradiated water. The system investigated contained one or two solutes. In this latter case, either one of the constituents was oxygen, or both were organic substances. The data obtained called attention to the complexity of the process induced by radiation. As was observed in solutions of inorganic substances, here also the yields of products were independent, over a considerable range of concentrations, of the substances irradiated, and a pronounced pH effect was found.

In the preceding review of the various studies carried out in the period between the world wars, we have mentioned only those which seemed to us to be important for giving a general picture. However, it should be noted that not a great many have been omitted. It seems that chemists were simply not very interested in chemical changes induced by radiations. As we shall see somewhat later, it was the development of nuclear energy during World War II that first called for more intense work in this branch of chemistry and also offered suitable facilities for it. Therefore, it is also not surprising to find that no particular efforts were made before 1942 to understand the essence of processes induced by radiation.

It is true, however, that two such attempts deserve attention. The first refers to the explanation of ferrous sulfate oxidation given by Risse [60]. He assumed that water is decomposed into H and OH radicals under the action of radiation. Recombinations of these species lead to the formation of H_2 and H_2O_2. According to Risse, it is hydrogen peroxide that oxidizes ferrous ions. The fact that in the presence of oxygen the oxidation yield is practically doubled only indicates that the concentration of hydrogen peroxide is also increased because of the oxygen. This increase is due to the reaction $2H + O_2 = H_2O_2$. At present, the mechanism of ferrous ion oxidation is well established (see Chapter Five, Section III). The assumptions made by Risse are certainly simpler than what we know at present actually occurs. Their great shortcoming is evidently the disregard of the OH radical in ferrous ion oxidation. However, Risse clearly pointed out what lies at the base of the actual explanation: the existence of free radicals and reaction of oxygen with H atoms.

The second attempt is Fricke's concept of activated water. The starting point was that X rays do not decompose pure, degassed water. Also, if certain substances are present in the water, then X-irradiation often leads to their chemical transformation, and the amount of changes induced is independent of the concentration of the solutes over a very wide range of concentration. From radiation physics it was known that most of the X-ray energy is absorbed by the water and not by the solutes which chemically change. Hence, Fricke concluded that X rays somehow activate water, and the activated water molecules must live a sufficiently long time to react with the solutes, especially when these are present in very low concentrations. In the absence of solutes, these activated water molecules must lose their energy in a reaction which, at low solute concentration, is in competition with their reaction with the solute. It should be noticed that by "activated water" was meant some sort of excited state of the water molecules that spontaneously decayed, and not species of free radicals. The rejection of free radicals was due to observations that dose rate (radiation intensity) has no effect on the yields observed. It should be mentioned that the effect of dose rate on the measured yield of radiolytic products was observed much later, when particle accelerators providing high beam-intensities became available.

It was not only the low intensity of natural radioactive sources that was an obstacle for their wider use in the radiation chemistry of aqueous solutions. Contamination was an even greater inconvenience. This held particularly in the case of studying the action of α rays, where the radiation source was dissolved in the aqueous system studied. Such studies became quite rare once X-ray sources became available. One of the later works was the paper of Nürnberger [61], published in 1934. He used α rays from radon dissolved directly in the solution or, in a procedure which the author calls indirect, contained in a glass ampul with thin walls which was dipped into the solution studied. The yields of the hydrogen and oxygen produced were found to be somewhat different from the data given by Cameron and Ramsay [35]. The oxidation of ferrous sulfate in 0.8 N H_2SO_4 was followed, and the amount of the hydrogen produced in this process was measured. It was found that the oxidation yield increased with increasing concentration of ferrous sulfate. Lanning and Lind [62] have reported data on the radiolysis of a number of solutions subjected to α rays from radon. They found that HI is decomposed into hydrogen and iodine, and HBr into hydrogen and bromine. In irradiated aqueous solutions of I_2, hydrogen iodide is formed. Potassium permanganate is reduced. The reaction yields were discussed in terms of the ratio M/N, where M is the number of converted species and N is the number of ion pairs produced in the system during the irradiation. The authors also concluded that the primary effect of radiation is the action on water, and that the changes observed are a consequence of reactions of the species produced by this action. They represented the primary effect and water decomposition products by the reaction $2H_2O + \text{energy} = H_2 + H_2O_2$.

V. FREE RADICALS AS THE CARRIERS OF CHANGES IN IRRADIATED WATER

We have seen that the first steps in radiation chemistry were associated with radiation sources made of natural radioactive substances, and that the realization of powerful X-ray devices enabled chemists to attain easily amounts of chemical change sufficient for more precise and reliable studies. Investigations in the field of nuclear energy in the United States during World War II were a new incitement to chemists. As a matter of fact, in the first case there was urgent need to know the behavior of water in a strong radiation field. Water, ordinary as well as heavy, was understood from the very beginning to be an important material for the construction of nuclear reactors, both as a moderator and as a cooler. The question arose as to how it would behave in a field of mixed radiation corresponding to that from 1000 kg radium. It was known that the main components of the mixed radiation are γ rays and fast neutrons. It was thought that the latter, through the intermediary or recoil protons that they produce in water, would give rise to changes similar to those caused by α particles. But, in contrast to photons, α particles were known to decompose water effectively. The cause of the difference between the chemical effects of photons and charged particles was not known, and it was not clear what was to be expected from a mixture of these radiations. Allen, one of the participants in this research, has written in this connection:

> Physicists had decided that the best scheme for cooling the reactor would be by water rushing through aluminum pipes. The difficulty was that no one knew whether under the influence of the unprecedentedly intense radiation in these reactors the water might react with aluminum as if it were calcium or sodium, as indeed thermodynamics suggested [7, p. 290]

Writing about the radiation chemistry program carried out under the guidance of Milton Burton and James Franck, he gives the following details:

> The thinking of Burton and Franck on the nature of activated water was based on the free radical theory, derived entirely from physical reasoning. Radiation could do nothing in the first instance except give energy to the electronic system of the water molecule, leading to ionization or to excited states similar to those produced by absorption of ultraviolet light. Now ionization would lead to water ion, H_2O^+, and a free electron. The latter, it was thought, would surely react with water to give the stable OH^- ion and thereby liberate a free hydrogen atom:

$$H_2O + e^- \longrightarrow OH^- + H \tag{1}$$

> The positive ion H_2O^+ would surely, in such a polar medium as water, give up a proton to the water to form very stable acid ion H_3O^+ and thereby liberate the radical OH:

$$H_2O^+ + H_2O \longrightarrow H_3O^+ + OH \tag{2}$$

> Excitation, on the other hand, should simply lead to the same state as formed by the absorption of light, which we have seen gives in water vapour only a continuous spectrum, thus indicating all the excited states to be dissociative and to lead to H + OH from the dissociation of a single molecule. So it seemed that these free radicals would be the only reasonable chemical intermediates to be postulated. The reasoning of Franck and Burton in this field and the experiments they indicated could not be published because the work was conducted under military secrecy [5, p. 14].

In the first papers published after the war, Burton [63] and Allen [64] summarized in the following way this radical model of water radiolysis: The ionization process

$$H_2O \;\rightsquigarrow\; H_2O^+ + e^- \tag{3}$$

V. FREE RADICALS AS THE CARRIERS OF CHANGES

is followed by reactions (1) and (2), opposite. Here it should be mentioned that the symbol ⤳ was introduced by Burton [63] to denote radiation-chemical reactions; this symbol in radiation chemistry has the same significance as $\xrightarrow{h\nu}$ in photochemistry.

In acid media the following reaction is also important:

$$H_3O^+ + e^- \longrightarrow H + H_2O \tag{4}$$

The fate of some of the radicals produced in reactions (1), (2), and (4) lies in the recombination reactions

$$H + H \longrightarrow H_2 \tag{5}$$

$$OH + OH \longrightarrow H_2O_2 \tag{6}$$

$$H + OH \longrightarrow H_2O \tag{7}$$

As to excited water molecules, which are also produced by radiation, it was said that some of them "might be expected to decompose into H and OH before being deactivated by collision." As regards the explanation of why α rays decompose carefully purified and degassed water and X rays do not, it is to be sought in the fact that the reactions

$$OH + H_2 \longrightarrow H_2O + H \tag{8}$$

$$H + H_2O_2 \longrightarrow H_2O + OH \tag{9}$$

lead to decomposition of the accumulated radiolysis products. This renders the equilibrium concentrations of H_2 and H_2O_2 in the case of X-radiation so low that these cannot be detected by the methods of analysis which are used in such cases. In the case of α-irradiation, the reversible reactions (8) and (9) are less efficient and hence the concentrations of the products are higher and more easily measurable. Irreproducibility of data on the equilibrium concentrations of hydrogen and hydrogen peroxide is accounted for by the chain character of the reactions—even the slightest concentrations of impurities react with the radical carrier of the chain reaction to give products which are incapable of carrying on the chain.

In the chapters to follow we shall see how little the radical model represented by reactions (1)–(9) has been modified in the course of 20 years. The significance of Allen's paper [64] also lies in a number of other considerations of the above reactions and of factors having effects on these. It was stated, among other things, that recombinations (5)–(7) have higher probability in the case of radiations having higher ionizing power. This affects the equilibrium concentrations of stable products and reactions (8) and (9). It was established that "almost anything dissolved in the water has the effect of increasing the amount of decomposition," and in this connection the effect of different solutes was shown. A study [65] published somewhat later, dealing with water decomposition under the action of mixed pile radiation (composed of fast neutrons and γ rays) confirmed the general picture of water radiolysis with the radicals H and OH and the molecular products H_2 and H_2O_2. The oxygen found in certain cases is not a primary product of water radiolysis, but arises from decomposition of the hydrogen peroxide.

The water radiolysis model described earlier, which was used in the United States, is certainly also due to the work of Weiss [66], reported in June of 1944. As a matter of fact, this is the first published work in which it was unequivocally stated that water decomposition under the action of ionizing radiation leads to the formation of H and OH radicals, and that the former are responsible for reductions in aqueous solutions (exemplified by $Ce^{4+} + H = Ce^{3+} + H^+$) and the latter for observed oxidations (exemplified by $Fe^{2+} + OH = Fe^{3+} + OH^-$). The presence of oxygen results in the occurrence of the reaction $H + O_2 = HO_2$, accompanied by the formation of hydrogen peroxide due to the reaction, $2HO_2 = H_2O_2 + O_2$. The decomposition is always followed by the recombination of radicals to give water ($OH + H = H_2O$) and $2H = H_2$. The result of the recombination of OH radicals, however, is seen by Weiss as $2OH = H_2O + O$, followed by $2O = O_2$. Quantitative relations were given from which it was clearly seen that observed changes of a substance S in solution are a consequence of the competition of free radicals in recombination reactions and of reaction with S. The "protection effect," known from radiobiological experiments, as well as the fact that observed changes are independent of the concentration of S, were explained. In the former case a certain added substance S_2 protects S_1 if it reacts with the same radical as S_1, provided that $k_2[S_2] > k_1[S_1]$. Here, k_2 and k_1 are the rate constants (in liters per mole per second) and $[S_2]$ and $[S_1]$ are the concentrations (in moles) of the substances studied.

As we have seen, in the first years following World War II it was generally accepted that water decomposition gives rise to free radicals having reducing (H) or oxidizing (OH) properties, as well as to the molecular products H_2 and H_2O_2. A basis was thereby provided for qualitative explanations of various phenomena observed in irradiated water and aqueous solutions. For a quantitative explanation it was necessary to know radiation-chemical yields for the free radicals and the rate constants for their reactions with one another or with various substances. Considerable progress in this direction was made in the subsequent years, as will be seen in later chapters. However, we shall conclude this section dealing with the free-radical model of water radiolysis by giving a concise review of some more important events that, in our view, are necessary to complete the general picture.

In 1952, a general discussion on radiation chemistry was organized by the Faraday Society and held in Leeds (Great Britain). It considered in great detail different aspects of radiation absorption, radiation dosimetry, radiation-

V. FREE RADICALS AS THE CARRIERS OF CHANGES

induced water decomposition, and chemical transformations in dilute aqueous solutions [67]. In the same year, Hochanadel [68] reported the first measurements of the radiation-chemical yields of free radicals considered as water radiolysis products, and Hart [69,70] used the measured yields of stable products in the radiolysis of oxygenated aqueous solutions of formic acid to calculate the yields of "primary reactions" at various pH's. Here, it should be noted that by the radiation-chemical yield is meant the number of species produced or disappearing per 100 eV of energy absorbed. This notion, corresponding in purpose to the ratio M/N in photochemistry, was first mentioned in a paper by Burton [63], where the symbol G was also introduced to represent this radiation-chemical yield:

$$G = \frac{\text{Number of species produced or disappearing}}{100 \text{ eV absorbed}}$$

The experimental yield of the product P formed is denoted by $G(P)$, whereas the yield of the substance S which disappears is denoted by $G(-S)$. One of the notations frequently used for the primary products of water radiolysis is G_R or G_M, where R = H, OH, e_{aq}^-, and M = H_2 or H_2O_2. Dainton and Sutton [71] showed that $G_{H_2} < G_{H_2O_2}$.

That the short-lived reactive species in water are formed only along the particle tracks (track effect) was pointed out as early as 1946 by Lea [72]. However, in 1953, Samuel and Magee [73,74] gave the first theoretical treatment of early processes along the tracks which give rise to the radical and molecular products. The essential assumption is that the electron ejected by the ionization of a water molecule cannot reach a sufficient distance from the parent ion and escape its coulombic field. The attraction leads to neutralization and excitation of the water molecule, which dissociates into H and OH spatially close to each other. This nonhomogeneous distribution of the primary species is time-limited as the radicals diffuse into the bulk of the solution from their place of origin. In the same year, Platzman [75] suggested that the electron may reach a sufficient distance from the parent ion and escape its coulombic field; after having been thermalized it becomes hydrated. Somewhat earlier, Day and Stein [76], and Stein [77] also assumed that electrons originating in ionization processes, when slowed down to energies at which they can interact with the solvent by polarization forces, may be directly captured by suitable solutes. Sworski [78] showed in 1954 that $G_{H_2O_2}$ may decrease if the concentration of scavenger for OH radicals is sufficiently increased. A year later, Schwarz [79] arrived at an analogous conclusion for G_{H_2}, which decreases in the presence of high concentrations of scavenger for the reducing species.

In 1955 and 1958, the Academy of Sciences of the USSR published collections of papers summarizing the results obtained at different Soviet laboratories; most of these contributions concern the radiolysis of water and aqueous solutions [*80,81*].

The discovery of the existence of two kinds of reducing species—the hydrogen atom and the hydrated electron—led to considerably more profound insight into the nature of short-lived reactive radiolysis products. This was the result of simultaneous efforts made at different laboratories. It was a gradually developing idea, as can be seen in the papers published by Baxendale and Hughes [*82*], and Hayon and Weiss [*83*] in 1958, those of Barr and Allen [*84*] in 1959, and those of Allan and Scholes [*85*], and Hayon and Allen [*86*].

In 1962, Czapski and Schwarz [*87*], and Collinson *et al.* [*88*] showed that the species supposed to be a hydrated electron, indeed has unit negative charge. In the same year, Hart and Boag [*89*], and somewhat later Boag and Hart [*90*], and Keene [*91*] reported the direct observation of the absorption spectrum of the hydrated electron in water.

VI. Pulse Radiolysis

Significant progress in the investigation of radiation-chemical reactions, as well as of rapid reactions in general, was made when the chemist was provided with a radiation source that allowed him to bring into the system studied, in one millionth of a second, an amount of energy sufficient to produce the reaction intermediates in sufficiently great concentrations for their detection by some physical method before they disappear. The possibility of depositing energy in a time short in comparison with the half-life of the reactive species observed, means a simplification of conditions for study of the kinetics of the process. The shortness of irradiation time also means a possibility of gaining insight into the earliest stages of the process and of observing many reactions for which, at the best, only assumptions could be made. These facilities were provided by pulsed electron beams in the 1960s. Their significance is not less than that of the facilities offered to photochemistry by flash photolysis in the 1950s. In fact, pulse radiolysis is in many respects the radiation-chemical analog of flash photolysis, and it is rather surprising that there was such a large time interval between these two realizations. True, as far back as 1954, Sangster [*92*] modified the Van de Graaff generator in such a way that it could carry out irradiations with 2.5-MeV electron pulses lasting 1 μsec and giving a peak current of 1 A. The object of these experiments was to produce a sufficient concentration of OH radicals in order that these might be detected spectrophotometrically. However, as shown by subsequent work, the problem chosen was too complex. In contrast, other investigators in the United States and England chose as the object of observation, species with a more suitable absorption spectrum. The results were reported almost simultaneously in 1960 by Matheson and Dorfman [*93*], McCarthy and MacLachlan [*94*], and Keene [*95*]. By the use of pulsed technique in radiation chemistry a great deal of new and varied information has been obtained within less than a decade [*19,20,26,27,29,96*]. This includes various observations of the hydrated electron, inorganic and organic radicals, and different radical ions. Pulsed radiolysis has enabled us to gain insight into the formation and behavior of cations with unusual valencies, such as Cu^+, Zn^+, Ni^+, Pb^+, Co^+, Cd^+, etc. The optical absorption spectra of numerous transient species have been reported [*97*]. The values of absolute rate constants for more than 1000 different reactions induced directly or indirectly by radiation in water and aqueous solutions have been established [*98*]. The extraordinary possibilities offered by the direct observation of the hydrated electron have up to now been among those most exploited. We shall see, however, that the contribution of pulse-radiolytic studies of other species to our understanding of the radiation chemistry of aqueous solutions is also considerable. All these results have been obtained mainly by experiments on the microsecond scale. However, the results of experiments where measurements were carried out in times of the order of magnitude of a nanosecond [*99*] or subnanosecond [*100*] have already begun to appear. New accelerators are in prospect which will give pulses of 10^{-10} sec or even shorter, while still providing enough energy in the pulse to produce a sufficiently high concentration of products to allow their investigation by optical absorption. This means that not only radiation-chemical measurements, but chemistry in general will enter a whole new dimension of time resolution.

At present, photochemistry and flash-photolysis studies are used increasingly to supplement radiation-chemical ones; the nature of some short-lived species can often be studied under simplified conditions. However, these fields are beyond the scope of the present book, and we refer the reader to papers of general character [*101–106*].

References

1. M. Burton, Radiation Chemistry. A Godfatherly Look at its History and Its Relation to Liquids. *Chem. Eng. News* Feb. 10, 86 (1969).
2. R. W. Clark, Selected Abstracts of Atomic Energy Project Unclassified Reported Literature in the Field of Radiation Chemistry and Bibliography of the Published Literature, AERE—C/R 1575 (1–6), Suppl. 2–5 (1956–1961).
3. Internal Report, Boris Kidrič Institute of Nuclear Sciences, Vinča, Yugoslavia, 1969.
4. E. J. Hart, Development of the Radiation Chemistry of Aqueous Solutions. *J. Chem. Educ.* **36**, 266 (1959).
5. A. O. Allen, Mechanism of the Radiolysis of Water by Gamma Rays or Electrons. *In* "Actions chimiques et biologiques des radiations" (M. Haissinsky, ed.), Vol. 5, p. 10. Masson, Paris, 1961.

6. A. O. Allen, Hugo Fricke and the Development of Radiation Chemistry: A Perspective View. *Radiat. Res.* **17**, 255 (1962).
7. A. O. Allen, Radiation Chemistry Today. *J. Chem. Educ.* **45**, 290 (1968).
8. A. O. Allen, "The Radiation Chemistry of Water and Aqueous Solutions." Van Nostrand–Reinhold, Princeton, New Jersey, 1961.
9. E. J. Hart and R. L. Platzman, Radiation Chemistry. *In* "Mechanisms in Radiobiology" (M. Errera and A. Forssberg, eds.). Academic Press, New York, 1961, Vol. 1, p. 93.
10. M. Haissinsky, ed., "Actions chimique et biologiques des radiations," Vols. 1–14 (1955–1970). Masson, Paris.
11. M. Ebert and A. Howard, eds., *Curr. Top. Radiat. Res.* **1–5** (1965–1969). North-Holland Publ., Amsterdam.
12. M. Burton and J. L. Magee, eds., *Advan. Radiat. Chem.* **1** (1969). Wiley (Interscience), New York.
13. M. Haissinsky, "La chimie nucléaire et ses applications." Masson, Paris, 1957.
14. E. Roth, "Chimie nucléaire appliquée." Masson, Paris, 1968.
15. A. J. Swallow, "Radiation Chemistry of Organic Compounds." Pergamon Press, Oxford, 1960.
16. I. V. Vereshchinskii and A. K. Pikaev, "Vedenie v radiatsionnuyu khimiyu." Akademiia Nauk SSSR, Moskva, 1963.
17. J. W. T. Spinks and R. J. Woods, "An Introduction to Radiation Chemistry." Wiley New York, 1964.
18. H. A. Schwarz, Recent Research on the Radiation Chemistry of Aqueous Solutions. *Advan. Radiat. Biol.* **1**, 1 (1964).
19. A. K. Pikaev, "Impulsnyi radioliz vody i vodnykh rastvorov." Nauka, Moskva, 1965.
20. M. S. Matheson and L. M. Dorfman, "Pulse Radiolysis." M. I. T. Press, Cambridge, Massachusetts, 1969.
21. V. N. Shubin and S. A. Kabakchi, "Teoriia i metody radiatsionoi khimii vody." Nauka, Moskva, 1969.
22. "Trudy I vsesoiuznogo soveshcheniia po radiatsionnoi khimii." Akademiia Nauk SSSR, Moskva, 1958.
23. "Trudy II vsesoiuznogo soveshcheniia po radiatsionnoi khimii." Akademiia Nauk SSSR, Moskva, 1962.
24. J. Dobó, ed., "Proceedings of the 1962 Tihany Symposium, Radiation Chemistry." Akadémiai Kiadó, Budapest, 1964.
25. J. Dobó and P. Hedvig, eds., "Proceedings of the Second Tihany Symposium on Radiation Chemistry." Akadémiai Kiadó, Budapest, 1967.
26. M. Ebert, J. P. Keene, A. J. Swallow, and J. H. Baxendale, eds., "Pulse Radiolysis." Academic Press, New York, 1965.
27. R. F. Gould, ed., "Solvated Electron" (*Adv. Chem. Ser. 50*). Am. Chem. Soc., Washington, D.C., 1965.
28. G. R. A. Johnson and G. Scholes, eds., "The Chemistry of Ionization and Excitation." Taylor and Francis, London, 1967.
29. R. F. Gould, ed., "Radiation Chemistry," Vol. 1 (*Advan. Chem. Ser. 81*). Am. Chem. Soc., Washington, D.C., 1968; "Radiation Chemistry," Vol II (*Advan. Chem. Ser. 82*). Am. Chem. Soc., Washington, D.C., 1968.
30. G. Stein, ed., "Radiation Chemistry of Aqueous Systems." Wiley, New York, 1968.
31. M. Ebert and A. Howard, eds., "Radiation Effects in Physics, Chemistry and Biology." North-Holland Publ., Amsterdam, 1963.
32. G. Silini, ed., "Radiation Research 1966." North-Holland Publ., Amsterdam, 1967.

33. F. Giesel, Über Radium und radioactive Stoffe. *Ber. Deut. Chem. Ges.* **35**, 3608 (1902); Über den Emanationskörper aus Pechblende und über Radium. *Ber. Deut. Chem. Ges.* **36**, 342 (1903).
34. W. Ramsay, The Chemical Action of the Radium Emanation; Part I. Action on Distilled Water. *J. Chem. Soc.* **91**, 931 (1907).
35. A. T. Cameron and W. Ramsay, The Chemical Action of Radium Emanation, Part III, On Water and Certain Gases. *J. Chem. Soc.* **92**, 966, (1908).
36. A. T. Cameron and W. Ramsay, The Chemical Action of Radium Emanation, Part IV, On Water. *J. Chem. Soc.* **92**, 992 (1908).
37. A. Debierne, Sur la décomposition de l'eau par les sels de radium, *C. R. Acad. Sci. Paris* **148**, 703 (1909).
38. M. Kernbaum, Action chimique sur l'eau des rayons pénétrants du radium, *C. R. Acad. Sci. Paris* **148**, 705 (1909).
39. M. Kernbaum, Action chimique sur l'eau des rayons pénétrants du radium, *C. R. Acad. Sci. Paris* **149**, 116 (1909).
40. W. Duane and O. Scheuer, Recherches sur la décomposition de l'eau par les rayons α. *Radium (Paris)* **10**, 33 (1913).
41. A. Kailan, Über die chemischen Wirkungen der durchdringenden Radiumstrahlung, 1. Der Einfluss der durchdringenden Strahlen auf Wasserstoffsuperoxyd in neutraler Lösung. *Wien. Ber.* **120**, 1213 (1911).
42. A. Kailan, Über die chemischen Wirkungen der durchdringenden Radiumstrahlung, 2. Der Einfluss der durchdringenden Strahlen auf Alkalijodide in wässeriger Lösung, *Wien. Ber.* **120**, 1373 (1911).
43. A. Kailan, Über die chemischen Wirkungen der durchdringenden Radiumstrahlung, 3. Der Einfluss der durchdringenden Strahlen auf einige anorganische Verbindungen, *Monatsh. Chem.* **33**, 1329 (1912).
44. A. Kailan, Über die chemischen Wirkungen der durchdringenden Radiumstrahlung, 4. Der Einfluss der durchdringenden Strahlen auf einige organische Verbindungen und Reaktionen, *Monatsh. Chem.* **33**, 1361 (1912).
45. A. Kailan, Über die chemischen Wirkungen der durchdringenden Radiumstrahlung, 7. *Monatsh. Chem.* **34**, 1269 (1913).
46. A. Kailan, Über die chemischen Wirkungen der durchdringenden Radiumstrahlung, 14. Die Einwirkung auf Oxalsäure, Kaliumtetraoxalat und Kaliumchlorat, *Monatsh. Chem.* **43**, 1 (1922).
47. A. Debierne, Recherches sur les gaz produits par les substances radioactives. Décomposition de l'eau. *Ann. Phys. (Paris)* **2**, 97 (1914).
48. H. Fricke and B. W. Petersen, Chemische, kolloidale und biologische Wirkungen von Röntgenstrahlen verschiedener Wellenlänge in ihrem Verhältnis zur Ionisation in Luft. I. Oxyhämoglobin in wässeriger Lösung. *Strahlentherapie* **26**, 329 (1927).
49. H. Fricke and S. Morse, The Action of X Rays on Ferrous Sulfate Solutions. *Phil. Mag.* **7**, 129 (1929).
50. N. Fricke and S. Morse, The Chemical Action of Roentgen Rays on Dilute Ferrous Sulfate Solutions as a Measure of Dose. *Amer. J. Roentgenol. Radium Ther.* **18**, 430 (1927).
51. O. Risse, Über die Röntgenphotolyse des Hydroperoxyds. *Z. Phys. Chem. Abt. A* **140**, 133 (1929).
52. H. Fricke and E. R. Brownscombe, Inability of X Rays to Decompose Water. *Phys. Rev.* **44**, 240 (1933).
53. H. Fricke, Reduction of Oxygen to Hydrogen Peroxide by the Irradiation of its Aqueous Solutions with X Rays. *J. Chem. Phys.* **2**, 349 (1934).

54. H. Fricke and E. J. Hart, The Oxidation of Fe^{2+} to Fe^{3+} by the Irradiation with X Rays of Solutions of Ferrous Sulfate in Sulfuric Acid. *J. Chem. Phys.* **3**, 60 (1935).
55. G. L. Clark and L. W. Pickett, Some New Experiments on the Chemical Effect of X Rays and the Energy Relations Involved. *J. Amer. Chem. Soc.* **52**, 465 (1930).
56. W. E. Roseveare, The X Rays Photochemical Reaction between Potassium Oxalate and Mercuric Chloride. *J. Amer. Chem. Soc.* **52**, 2612 (1930).
57. G. L. Clark and W. S. Coe, Photochemical Reduction with X Rays and Effects of Addition Agents. *J. Chem. Phys.* **5**, 97 (1937).
58. H. Fricke and E. R. Brownscombe, The Reduction of Chromate Solutions by X Rays. *J. Amer. Chem. Soc.* **55**, 2358 (1933).
59. H. Fricke, E. J. Hart, and H. P. Smith, Chemical Reactions of Organic Compounds With X Rays Activated Water. *J. Chem. Phys.* **6**, 229 (1938).
60. O. Risse, Einige Bemerkungen zum Mechanismus chemischer Röntgenreaktionen in wässerigen Lösungen, *Strahlentherapie* **34**, 578 (1929).
61. C. E. Nurnberger, Effect of Alpha Particles on Aqueous Solutions, I. The Decomposition of Water. II. The Oxidation of Ferrous Sulfate. *J. Phys. Chem.* **38**, 47 (1934).
62. F. C. Lanning and S. C. Lind, Chemical Action of Alpha Particles from Radon on Aqueous Solutions. *J. Phys. Chem.* **42**, 1229 (1938).
63. M. Burton, Radiation Chemistry. *J. Phys. Colloid Chem.* **51**, 611 (1947).
64. A. O. Allen, Radiation Chemistry of Aqueous Solutions. *J. Phys. Colloid Chem.* **52**, 479 (1948).
65. A. O. Allen, C. J. Hochanadel, J. A. Ghormley, and T. W. Davis, Decomposition of Water and Aqueous Solutions under Mixed Fast Neutron and Gamma Radiation. *J. Phys. Chem.* **56**, 575 (1952).
66. J. Weiss, Radiochemistry of Aqueous Solutions. *Nature* **153**, 748 (1944).
67. "Radiation Chemistry." *Discuss. Faraday Soc.* **12** (1952).
68. C. J. Hochanadel, Effects of Cobalt γ-Radiation on Water and Aqueous Solutions. *J. Phys. Chem.* **56**, 587 (1952).
69. E. J. Hart, The Radical Pair Yield of Ionizing Radiation in Aqueous Solutions of Formic Acid. *J. Phys. Chem.* **56**, 594 (1952).
70. E. J. Hart, γ-Ray Induced Oxidation of Aqueous Formic Acid-Oxygen Solutions. Effect of pH. *J. Amer. Chem. Soc.* **76**, 4198 (1954).
71. F. S. Dainton and H. S. Sutton, Hydrogen Peroxide Formation in the Oxidation of Dilute Aqueous Solutions of Ferrous Sulfate by Ionizing Radiation. *Trans. Faraday Soc.* **49**, 1011 (1953).
72. D. E. Lea, "Action of Radiation on Living Cells." Cambridge Univ. Press, London and New York, 1946.
73. A. H. Samuel and J. L. Magee, Theory of Radiation Chemistry. II. Track Effects in Radiolysis of Water. *J. Chem. Phys.* **21**, 1080 (1953).
74. J. L. Magee, Mechanisms of Energy Degradation and Chemical Change Effects of Secondary Electrons. *In* "Basic Mechanisms in Radiobiology. II. Physical and Chemical Aspects" (J. L. Magee, M. D. Kamen, and R. L. Platzman, eds.), p. 51. National Research Council, Publ. No. 305, Washington, D.C. 1953.
75. R. L. Platzman, Energy Transfer from Secondary Electrons to Matter. *In* "Basic Mechanisms in Radiobiology. II. Physical and Chemical Aspects" (J. L. Magee, M. D. Kamen, and R. L. Platzman, eds.), p. 22. National Research Council, Publ. No. 305, Washington, 1953.
76. M. J. Day and G. Stein, Chemical Dosimetry of Ionizing Radiations. *Nucleonics* **8**, No. 2, 34 (1951).

77. G. Stein, Some Aspects of the Radiation Chemistry of Organic Solutes. *Discuss. Faraday Soc.* **12**, 227 (1952).
78. T. J. Sworski, Yields of Hydrogen Peroxide in the Decomposition of Water by Cobalt γ-Radiation. I. Effect of Bromide Ion. *J. Amer. Chem. Soc.* **76**, 4687 (1954).
79. H. A. Schwarz, The Effect of Solutes on the Molecular Yields in the Radiolysis of Aqueous Solutions. *J. Amer. Chem. Soc.* **77**, 4960 (1955).
80. "Sbornik rabot po radiatsionnoi khimii." Akademiia Nauk SSSR, Moskva, 1955.
81. S. Ia. Pshezhetskii, ed., "Deistvie ioniziruyshchikh izluchenii na neorganicheskie i organicheskie sistemy." Akademiia Nauk SSSR, Moskva, 1958.
82. J. H. Baxendale and G. Hughes, The X-Irradiation of Aqueous Methanol Solutions. Part I. Reactions in H_2O. Part II. Reactions in D_2O. *Z. Phys. Chem. (Frankfurt)* **14**, 306 (1958).
83. D. Armstrong, E. Collinson, F. S. Dainton, D. M. Donaldson, E. Hayon, N. Miller, and J. Weiss, Primary Products in the Irradiation of Aqueous Solutions with X or Gamma Rays. *Proc. Int. Conf. Peaceful Uses At. Energy, 2nd, Geneva, 1958* **29**, P/1517, 80 (1958).
84. N. F. Barr and A. O. Allen, Hydrogen Atoms in the Radiolysis of Water. *J. Phys. Chem.* **63**, 928 (1959).
85. J. T. Allan and G. Scholes, Effect of pH and the Nature of the Primary Species in the Radiolysis of Aqueous Solutions. *Nature* **187**, 218 (1960).
86. E. Hayon and A. O. Allen, Evidence for two Kinds of "H Atoms" in the Radiation Chemistry of Water. *J. Phys. Chem.* **65**, 2181 (1961).
87. G. Czapski and H. A. Schwarz, The Nature of the Reducing Radical in Water Radiolysis. *J. Phys. Chem.* **66**, 471 (1962).
88. E. Collinson, F. S. Dainton, D. R. Smith, and S. Tazuke, Evidence for the Unit Negative Charge on the "Hydrogen Atom" formed by the Action of Ionizing Radiation on Aqueous Systems. *Proc. Chem. Soc.* 140 (1962).
89. E. J. Hart and J. W. Boag, Absorption Spectrum of the Hydrated Electron in Water and in Aqueous Solutions. *J. Amer. Chem. Soc.* **84**, 4090 (1962).
90. J. W. Boag and E. J. Hart, Absorption Spectra of "Hydrated" Electron. *Nature* **197**, 45 (1963).
91. J. P. Keene, Optical Absorption in Irradiated Water. *Nature* **197**, 47 (1963).
92. M. Sangster, General Discussion. *Discuss. Faraday Soc.* **17**, 112 (1954).
93. M. S. Matheson and L. M. Dorfman, Detection of Short-Lived Transients in Radiation Chemistry. *J. Chem. Phys.* **32**, 1870 (1960).
94. R. L. McCarthy and A. MacLachlan, Transient Benzyl Radical Reactions Produced by High-Energy Radiation. *Trans. Faraday Soc.* **56**, 1187 (1960).
95. J. P. Keene, Kinetics of Radiation-Induced Chemical Reactions. *Nature* **188**, 843 (1960).
96. H. Greenshields and W. A. Seddon, Pulse Radiolysis a Comprehensive Bibliography (1960–March 1969), AECL-3524, Chalk River, 1970.
97. A. Habersbergerova, I. Janovski, and J. Teply, Absorption Spectra of Intermediates Formed during Radiolysis and Photolysis. *Radiat. Res. Rev.* **1**, 109 (1968).
98. M. Anbar and P. Neta, A Compilation of Specific Bimolecular Rate Constants for the Reactions of Hydrated Electrons, Hydrogen Atoms and Hydroxyl Radicals with Inorganic and Organic Compounds in Aqueous Solutions. *Int. J. Appl. Radiat. Isotopes* **18**, 493 (1967).
99. J. K. Thomas and R. V. Bensasson, Direct Observation of Regions of High Ion and Radical Concentration in the Radiolysis of Water and Ethanol. *J. Chem. Phys.* **46**, 4147 (1967).

100. M. J. Bronskill and J. W. Hunt, A Pulse-Radiolysis System for the Observation of Short-Lived Transients. *J. Phys. Chem.* **72**, 3762 (1968).
101. M. S. Matheson, Photochemical Contributions to Radiation Chemistry. *Proc. Int. Conf. Peaceful Uses At. Energy, Geneva, 2nd, 1958* **29**, P/949, 385 (1958).
102. G. Porter, Application of Flash Photolysis in Irradiation Studies. *Radiat. Res. Suppl.* **1**, 479 (1959).
103. G. Stein, Excitation and Ionization. Some Correlations Between the Photo and Radiation Chemistry of Liquids. *In* "The Chemistry of Ionization and Excitation" (G. R. A. Johnson and G. Scholes, eds.), p. 25. Taylor and Francis, London, 1967.
104. E. F. Caldin, "Fast Reactions in Solution," p. 104. Blackwell Scientific Publ., Oxford, 1964.
105. G. Porter, Flash Photolysis and Primary Processes in the Excited State. *In* "Fast Reactions and Primary Processes in Chemical Kinetics" (Stig Claesson, ed.), p. 141. Wiley (Interscience), New York, 1967.
106. U.S. Army Natick Laboratory, April 22–24, 1968, Symposium on Photochemistry and Radiation Chemistry. *J. Phys. Chem.* **72**, 3709–3929 (1968).

The passage of radiation through water gives rise to ions, secondary electrons, and excited water molecules. The time in which these species are produced is in fact the interaction time of radiation with a water molecule; this time depends on the type and energy of the radiation and ranges from 10^{-16} to 10^{-14} sec.

The lifetime of the species produced is very short, and in about 10^{-11} sec after the passage of radiation, thermal equilibrium is established in the water. During this time, the secondary electron is also thermalized. Various aspects of absorption of radiation are studied in radiation physics, and we refer the reader who is particularly interested in the radiation-chemical aspects to the relevant literature [1-4]. The excellent article of Mozumder [5] presents details about charged particle tracks and the track model in certain areas of radiation chemistry. Also, the proceedings of a recent symposium [6] summarize various aspects of very early effects. Section I of this chapter is a review of basic facts on the absorption of radiation in water, as Section II gives only some important data on excited and ionized water molecules. Further details should be sought in the specialized literature, in particular that dealing with optical, mass, electron-impact, and ESR spectra. Unfortunately, apart from a few exceptions, these data refer not to liquid water, which is the subject of the present book, but to its gaseous or solid state.

It seems generally accepted that the active species produced in the primary act of interaction of radiation with water—ionized and excited water molecules—do not take part in the chemistry that occurs in irradiated water or aqueous solution, because of their distribution and lifetime (Sections III and IV). This is why the primary agents of chemical changes are considered to be the products of their interactions—free radicals which are found in the water after thermal equilibrium has been established. On the basis of our present knowledge, these are the following relatively short-lived reactive species: the hydroxyl radical (OH), the hydrogen atom (H), and the hydrated electron (e_{aq}^-). At present, these species (not the species preceding them, H_2O^+, e^-, and H_2O^*) are considered to be the primary agents of chemical changes; they are generally called the primary products of water radiolysis. Here also H_2O_2 and H_2 are often added; they are formed slightly later in recombination reactions of corresponding radicals or, at least partially, simultaneously with them but through mechanisms which are unknown. The nature of these primary species, as they are often called, is considered in detail in Chapters Three through Five. The purpose of this chapter is to expose assumptions as to how they are generated in irradiated water. Since the theoretical information on the subject is rather scarce and the technical approach is limited, speaking of the origin of primary species amounts only to a discussion of possible reactions leading to their formation, i.e., to the consideration of a water radiolysis model. This latter has to take into account the available data on the chemical properties of irradiated water and is liable to corrections as further developments of theory and improvements of experimental facilities occur.

CHAPTER TWO

Interaction of Ionizing Radiation with Water and the Origin of Short-Lived Species That Cause Chemical Changes in Irradiated Water

I. Absorption of Radiation in Water

A. Gamma Rays and X Rays

High-energy photons lose a considerable part of their energy in every elementary act of interaction with the atoms of the medium they traverse. The way in which they lose energy in water depends mainly on the energy of the incident photon. When $E < 0.01$ MeV, the dominant process is the photoelectric effect, in which the energy of the incident photon is entirely transferred to the electron ejected from the atom encountered. If the photon in collision does not transfer all its energy to the electron, but only a part of it, which allows it to continue its travel in a deviated path, then the process is known as the Compton effect (after its discoverer). In practice, the absorption in water of γ or X rays of an energy from a few thousand to a few million electron volts proceeds predominantly by this effect. The third process, which is of far less importance under normal radiation-chemical conditions, is pair production. Pair production can occur when the energy of the photon exceeds 1 MeV, but it becomes important only at much higher energies. In this case the photon is completely absorbed near the nucleus, and its disappearance is accompanied by the production of a pair of charged particles—an electron and a positron.

At sufficiently high photon energies these processes take place simultaneously in a given medium, their relative abundance being dependent on the energy of the incident radiation and on the atomic number of the constituent atoms of the medium. However, it should be noted that every incident photon is not necessarily involved in one of these processes. The probability that a photon will interact during its passage depends on several factors related by the Lambert law of absorption:

$$I = I_0 e^{-\mu x} \tag{1}$$

Here I_0 is the intensity of the incident radiation (expressed in ergs per square centimeter per second), I is the intensity of radiation transmitted through a layer of thickness x of the absorber (in centimeters), and μ is the total linear absorption coefficient (expressed in reciprocal centimeters):

$$\mu_{tot} = \mu_{ph} + \mu_C + \mu_p \tag{2}$$

where μ_{ph}, μ_C, and μ_p are the absorption coefficients for photoeffect, the Compton effect, and pair production, respectively. Substituting the data on μ_{tot} into the equation representing the Lambert law, we can easily calculate the thickness of the water layer x (in centimeters) necessary for the desired percentage of incident radiation to be absorbed. We can see that electromagnetic radiation of an energy of 1 MeV traverses several centimeters of water, losing only a few percent of its intensity, whereas soft X rays with an energy of a few thousand electron volts are practically completely absorbed in a few millimeters.

The Lambert law of absorption holds only for a well-focused beam of monoenergetic radiation. However, in radiation chemistry the situation is much more complex in practice. Not only is the incident beam insufficiently focused and polychromatic, but also the nature of the process of degradation of energy contributes to the complexity of the radiation energy spectrum. For example, in the Compton process both the scattered photon and the ejected electron have enough energy to cause new processes as they pass through water, provided the thickness of the water layer is sufficient. In the photoeffect, we have the ejected electron and secondary X-radiation (fluorescent radiation) from the atom that lost the electron. Annihilation of the positron generated in the process of pair production also gives rise to a new photon. A consequence of all this is that energy transfer in irradiated water is much more complex than it appears from the Lambert law of absorption taken alone.

Although the Lambert law allows only a calculation of the decrease in intensity of the incident beam, under certain conditions it can be used to calculate the amount of energy received by the sample irradiated. For this purpose, various calculations have been worked out [7], but in practice the dosimetric measurement is preferred, since it is more accurate and expedient (Chapter Eight).

Figure 2.1 illustrates the part of individual processes in the degradation of energy of the photon as it passes through water. It shows atomic absorption coefficients for individual processes as a function of the incident radiation

I. ABSORPTION OF RADIATION IN WATER

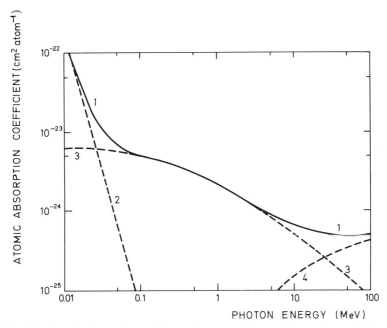

Fig. 2.1 Atomic absorption coefficients for water: the total absorption coefficient (1), the photoelectric coefficient (2), the total Compton coefficient (3), and the pair production coefficient (4).

energy. The atomic absorption coefficient is related to the linear absorption coefficient in the following manner:

$$\mu_{(A)} = \frac{\mu A}{\rho N_0} \quad \text{cm}^2 \text{ atom}^{-1} \tag{3}$$

where

- $\mu_{(A)}$ is the atomic absorption coefficient, square centimeters per atom;
- μ is the linear absorption coefficient, reciprocal centimeters;
- A is the atomic weight, grams per mole;
- ρ is the density, grams per cubic centimeter;
- N_0 is Avogadro's number, atoms per mole.

The atomic absorption coefficient has the dimensions of area, and is often referred to as the cross section. It can be seen from the figure that the Compton effect is the dominant process during the energy deposition in water, since in radiation chemistry, use is most often made of γ-radiation of an energy of the order of a few million electron volts. The Compton electrons have a mean energy of 440 keV in the case of γ rays from radioactive ^{60}Co.

B. Electrons

In contrast to the photon, the energy of which may decrease considerably in every act of interaction with the molecules of the medium it traverses, the electron loses its energy very gradually in ionizing and exciting a large number of molecules. In the interaction of a fast electron with a water molecule, an electron is often ejected and the incident electron continues its travel with a smaller energy on account of the ejected electron. If a very fast electron happens to pass by an atomic nucleus, which is rather an infrequent case under conditions in which radiation-chemical experiments are carried out (low Z of the medium and electron energies mostly less than 5 MeV), then the electron may suddenly be decelerated. The corresponding energy loss is manifested in the form of bremsstrahlung. As to the fate of low-energy incident electrons or secondary electrons produced in one of the processes mentioned previously, they chiefly lose their energy in the following ways: They excite molecules, neutralize the positive ions present, attach themselves to neutral molecules, or polarize the medium and become hydrated. Capture of an electron by a positive ion or by a neutral molecule may be accompanied by dissociation of the product obtained. It is convenient to classify the electrons in the following groups:

- Fast, with $E \gtrsim 100$ eV, where E is the kinetic energy of the electron. They predominantly excite and ionize to optically allowed states and occasionally produce fast secondary δ electrons in hard collisions.
- Slow, with $100 \text{ eV} > E > E_0$, where E_0 is the threshold of electronic excitation. These electrons excite and ionize to optically allowed states, but, by electron exchange mechanisms, they also ionize at lower energies to optically forbidden and to spin-forbidden states.
- Subexcitation, with $E_0 > E > kT$. They cause vibrational and rotational excitations, and, possibly, the neutralization of positive ions. Indirect excitation via temporary negative ion states is very important.
- Thermal or near thermal, with $E \sim kT$. Their fate is in attachment to neutral molecules (dissociative or nondissociative), in the charge neutralization processes, and in the interaction with the environment (e.g., by polarization).

The energy loss of an electron under given conditions can be calculated using the expression given by Bethe and Ashkin [1]. For experiments in radiation chemistry, we are mainly interested in the expression which takes into account energy loss in inelastic collisions, the dominating process for the electrons with $E \gtrsim 100$ eV,

$$-\frac{dE}{dx} = \frac{2\pi N e^4 Z}{m_0 v^2} \left[\log \frac{m_0 v^2 E}{2I^2(1-\beta^2)} - (2\sqrt{1-\beta^2} - 1 + \beta^2) \log 2 \right.$$
$$\left. + 1 - \beta^2 + \frac{1}{8}(1 - \sqrt{1-\beta^2})^2 \right] \quad (4)$$

Here,

- $-dE/dx$ is the energy loss per unit path, which is also called stopping power, ergs per centimeter,
- N is the number of atoms per cubic centimeter,
- e is the electron charge,
- Z is the atomic number of the stopping material,
- m_0 is the electron rest mass,
- E is the kinetic energy of the electron, ergs,
- v is the velocity of the electron, centimeters per second,
- β equals v/c, where c is the velocity of light,
- I is the average excitation potential of the atom of the stopping material, ergs, which is often called the stopping potential and expressed in electron volts.

The value of I for water amounts to 66 eV [8] and is calculated from values given for the O and H atoms [9]. Taking into account that $I = 66$ eV, as well as the fact that no relativistic correction is needed when $E < 50$ keV, Allen [10] gives the simple expression for calculation of the energy loss of electrons in this case:

$$-\frac{dE}{dx} = \frac{1.019}{E} \log \frac{E}{56.6}. \quad (5)$$

Here, E is given in electron volts and $-dE/dx$ in electron volts per angstrom.

Using expression (4) or (5) for energy loss of electrons in water, we can see that $-dE/dx$ falls from about 0.22 eV Å^{-1} for electrons with an initial energy of 10 keV down to 0.02 eV Å^{-1} for $E = 400$ keV; it then remains practically constant as E continues to rise. As a matter of fact, at first it slowly decreases with further increase of E and introduction of the relativistic correction, attains a minimum value of 0.018 eV Å^{-1} for electron energies between 1 and 2 MeV, and then slowly rises again to attain the value 0.02 eV Å^{-1} [10] at 5 MeV.

C. Heavy Charged Particles

Protons, deuterons, helions, or α particles transfer their energy to the medium in the same way as electrons, ionizing and exciting water molecules in a number of successive collisions. This proceeds until the charged particles are so much slowed down that they may be neutralized by reaction with electrons, whereby their travel is stopped.

In view of the conditions under which experiments are carried out in studying the radiation chemistry of aqueous solutions, only the inelastic collisions of charged particles are of practical importance. As in the case of electrons [Eq. (4)], energy loss of these particles is given by the relation [1]

$$-\frac{dE}{dx} = \frac{4\pi Z^2 e^4}{m_0 v^2} NZ \left[\log \frac{2m_0 v^2}{I(1-\beta^2)} - \beta^2 \right] \quad (6)$$

Here, $-dE/dx$ is expressed in ergs per centimeter, Z is the charge on the particle, while all other quantities have the same meanings as in Eq. (4).

The above expression shows that the slower the particle and the higher the magnitude of its charge, the larger is the energy loss. If we compare two particles of the same energy but of different mass and charge, for example a deuteron and a helion, the latter will have a smaller velocity and a larger energy loss. Thus, we can see that, as the initial energy of a helion in water decreases from 20 to 5 MeV, $-dE/dx$ increases from about 3.3 to 9.5 eV Å^{-1}; the corresponding values for a deuteron are 0.48 and 1.42 eV Å^{-1} [10].

D. Neutrons

In contrast to the case of the radiations considered in Sections IA–C, the primary interaction of neutrons with a medium involves not the electron shells but the nuclei of the atoms. Neutrons with energies larger than that of thermal motion give up their energy in elastic collisions with nuclei. The smaller the mass of the nucleus encountered, the larger the energy loss. In water, such nuclei are those of hydrogen atoms which take up most of the energy of the colliding neutrons. In an elastic collision, the H atom takes up, on the average, about half the energy of the neutron, loses its electron, and begins to move as a proton. So, just as the interaction of a photon with an atom of medium gives rise to an electron, the interaction of a neutron with an H atom gives rise to a proton. This last reacts with the atoms of the medium as described in the previous section.

Thermal neutrons are most often captured by the nucleus with which they collide. In water, this leads to the nuclear reaction ^{1}H(n, γ)^{2}H, where the disappearance of the neutron is accompanied by the emission of a γ quantum with an energy of 2.2 MeV. The corresponding reaction with oxygen is practically negligible.

II. EXCITED AND IONIZED WATER MOLECULES

In the preceding section we have considered what happens with incident radiation and not with the water through which it passes. We have seen that energy loss in the case of charged particles is due to the Coulomb interaction between the electron shell of the struck atom and the incident particle. Depending on how close the ionizing particle passes and the velocity of its motion, the electron in the shell of the water molecule takes up a larger or a smaller amount of energy and momentum. A consequence of this energy and momentum transfer is a change in position and motion of the electrons. In stable molecules in the ground state, electrons are normally coupled into pairs with opposite spins. Excitation into an excited singlet state leads to the quantum transition of an electron to a higher energy level, but in such a way that the spin orientation does not change. If it does, one speaks about an excited triplet state; it is very important for radiation-chemical processes because of its biradical character and relatively long lifetime in condensed media. According to Stein [11,12], an excited triplet state of the water molecule is not likely to be effective in liquid-state radiolysis. Also, there is no convincing experimental evidence for its existence. However, some evidence for the triplet state in ice at 77°K has been reported [13].

If the energy taken up by the electron is larger than the ionization potential, it may part from the molecule leaving a positive ion. The formation and behavior of H_2O^+ and H_2O^* are fairly well established in the case where the H_2O molecule is sufficiently isolated, so that dissociation and detection of the fragments may take place before the active species collides with a neighboring molecule. This is the case in experiments carried out in the gas phase at low pressure (10^{-4} mm). At higher pressures or in liquid water, the situation is more complex due to secondary reactions and interference of other phenomena, such as the cage effect, diffusion, and hydration. This is why it is not surprising that we still have only fragmentary answers to the questions as to how, and to what extent, the excited and ionized water molecules contribute to the formation of the primary products of water radiolysis. In this connection, the papers of Fiquet-Fayard [14a,b], Platzman [2], Stein [11,12], and Santar and Bednář [15a–c] may be of interest for the reader.

A useful general picture may be based on the following reliable experimental data: electronic excitation threshold 6.6 eV [16], ionization potential $I = 12.6$ eV [17], and $W = 31$ eV [18]. This last value represents the observed "mean energy expenditure per ion-pair formation" and takes into account not only the absorbed energy which is consumed in ionizing processes but also that dissipated in all other ways.

It is experimentally established that in the range of normal excitation, that is, for absorbed energy below the ionization potential, three regions can be distinguished. The lowest, a band of truly continuous absorption with a maximum at 7.4 eV, is responsible for the well-known opacity of water at wavelengths shorter than about 1800 Å. This excitation leads to dissociation into H and OH in their electronic ground states. The other excitation regions include both continuous absorption and a number of sharp bands with maxima at 10 and 11 eV. In these cases, the dissociation products are H + OH* or H_2 and O.

Superexcitation is a very important process, since it contributes to more than half of the nonionizing events. The range of superexcitation extends from 12.6 to about 30 eV. It leads either to the ionization or to the dissociation of the water molecule, where an abundant formation of H* seems very probable. Different aspects of superexcited states of molecules are considered by Platzman [19].

The ionization leads to the formation of the H_2O^+ ion as by far the most abundant species. Mass-spectrometric data show that the abundance of OH^+ is four times lower, while those of H^+ and O^+ are lower by a factor of 20 and 50, respectively. It is interesting to note that H_2O^- is not detected and that other negative ions (O^-, OH^-, etc.) are present only in trace amounts. The values of appearance potentials and the identities of species involved as given by different authors vary slightly. Thus, we can find [14a] H_2O^+, 12.6–16.3 eV; OH^+, 18.1–18.7 eV; H^+, 19.5 eV; O^+, 28.1–29.1 eV; and O^-, 36 eV.

Some experimental facts about excited and ionized water molecules, mentioned in this section, also can be seen from a theoretical approach. Figure 2.2 is an approximation to the complex optical spectrum of H_2O [2]. The abscissa represents the transition energy in electron volts and the ordinate shows the abundance of a particular excitation or ionization state of isolated molecules, excited by photons of particular energies E expressed as df/dE.

If E_s is the energy of a molecule in the excited state s relative to its energy in the ground state (the excitation energy), then this abundance is given by f_s, the oscillator strength for the transition from the ground state to the discrete excited state s, induced by the absorption of light of frequency $v_s = E_s/h$. However, for transitions within a continuum, this discrete oscillator strength f_s must be replaced by the differential oscillator strength per unit energy range, df/dE. The relative abundance of excited and ionized states induced by fast charged particles (typically electrons) is now given by $(1/E)(df/dE)$, the derived excitation spectrum that is closely related to but different from the above-mentioned optical spectrum.

The starting point in these considerations is the rule called optical approximation [2,19,20]. It stems from Bethe's theory of individual inelastic collisions of charged particles with atoms and molecules. It is fully valid when applied to the action of fast electrons ($E \gtrsim 100$ eV) but progressively failing for slower

II. EXCITED AND IONIZED WATER MOLECULES

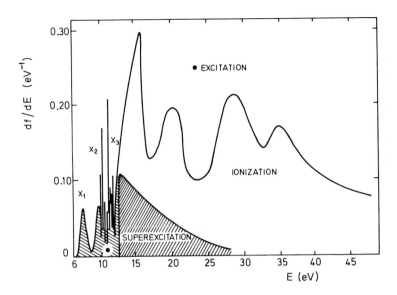

Fig. 2.2. Optical spectrum of water. (After R. L. Platzman [2]. Reproduced by permission of North-Holland Publishing Company.)

electrons ($E < 100$ eV). This rule was quite successfully applied in the theoretical calculations of the primary radiation-chemical yields. Two somewhat different approaches should be mentioned.

Platzman [2] has calculated the absolute yields of excitations and ionizations in the formation of active species of water subjected to ionizing radiation. A quantity of interest to the radiation chemist is the "ratio of excitation to ionization," for which a value of 0.42 is derived. An important conclusion was also that, in the first approximation, excitation spectra and the yields calculated in the optical approximation for the gaseous phase remain valid for condensed matter. The proximity of other molecules is likely to affect the further fate of the active species formed and perhaps to alter drastically the chemical reaction sequence or sequences which they initiate.

Santar and Bednář [15,21a–c] put more emphasis on the structure of ionizing particle tracks and on the contribution of slow secondary electrons. Table 2.1 gives a comparison of their findings with those mentioned above. The ratio of excitation to ionization is 0.97. The difference in the two sets of overall results derives from the separate treatment of slow electrons; the agreement is good when the two values are obtained from the excitation spectra. It is important to note that these authors conclude that the ratio of excitation to ionization should be markedly influenced by the molecular nature

TABLE 2.1

CALCULATED YIELD OF PRIMARY ACTIVATIONS IN WATER

Radiation	Excitations			Ionizations	Total	$\dfrac{\text{Excitations}}{\text{Ionizations}}$	Ref.
	Below I	Above I	Total				
Fast electrons	0.35	0.69	1.04	2.66	3.70	0.39	[15a,c]
Slow electrons	≳0.75	≲1.45	2.2 ± 1	0.66	2.8 ± 1	3.3 ± 1.5	[15a,c]
Total	≳1.10 ± 0.3	≲2.14 ± 0.7	3.2 ± 1	3.32	6.5 ± 1	0.97 ± 0.3	[15a,c]
Total	0.54	0.92	1.46	3.48	4.94	0.42	[2]

of the medium. Also, in contrast to Platzman's approach where a considerable part of the observed decomposition is implicitly expected to occur in subsequent physicochemical and chemical stages, they explain most of the observed decomposition by the processes of the physical stage.

According to some experimental evidence [22], in the radiolysis of water vapor free-radicals are produced in considerably lower abundance than the positive ions. The ratio 0.47 can be calculated from the experimental data.

The low-energy tail of the absorption spectrum of liquid water [23,24] is not basically different from that of water vapor, and this statement should be even more valid at higher energy transitions. Similar absorption bands, somewhat red-shifted, are observed below 2000 Å. The photolysis of liquid water at 1849 and 1470 Å indicates the formation of a radical pair [25-27] as the main primary process in both cases:

$$H_2O \xrightarrow{h\nu} H_2O^* \longrightarrow H + OH \qquad (7)$$

Owing to the cage effect, a considerable number of geminate pairs (H + OH) recombine to give water. The recombination is prevented by the presence of scavengers in water which compete for the free radicals produced in reaction (7). There is no strong experimental evidence that scavengers at concentrations less than 1 M directly react with water molecules in the primary excited state. The dependence of the radical yields on scavenger concentration is in accord with diffusion-controlled competition kinetics; a large fraction of molecular hydrogen formed in water radiolysis does not come directly from the dissociation of the water molecules. Experiments with water vapor at 1236 Å confirm the presence of molecular hydrogen [28,29]. Its yield depends [28] on the water vapor pressure and the presence of an efficient hydrogen scavenger, indicating the reaction

$$H + H_2O \longrightarrow H_2 + OH \qquad (8)$$

The same work also confirms the absence of reaction (8) at 1470 Å. It should be noted that the occurrence of reaction (8) cannot be completely ruled out in the photolysis of liquid water at 1236 Å, but no experimental evidence has as yet been reported for this wavelength. The photolysis of D_2O closely parallels that of light water [30]. The presence of molecular hydrogen was also explained [29] by means of the reaction $H_2O \rightarrow H_2 + O$, occurring with a probability three times lower than reaction (7).

The quantum yields of the products formed in reaction (7) during the photolysis of liquid water at 1236, 1470, and 1849 Å have been reported [31]. Formate and carbon dioxide were used as scavengers. The hydrated electron was also suggested as a probable product. Its yield increased with decreasing wavelength, similarly to the quantum yields of the H and OH species. However, it was about ten times lower than the latter. A direct observation of e_{aq}^- during the flash photolysis of highly purified water has been reported [32]. This was done by means of optical absorption measurements in irradiated solutions: The hydrated electron was identified by its absorption spectrum, which was known from pulsed electron beam experiments, as well as by its reactivity with O_2, H_3O^+, and N_2O. The quantum yield is evaluated to be about twenty times lower than the yields of H and OH in the steady-state photolysis of water. It is interesting to note that the wavelength limit for formation of the hydrated electron was found to be about 1920 Å, indicating that the "ionization potential" of liquid water is about 6.5 eV, compared to 12.6 eV for the vapor. Absorption spectra of the OH and H species produced in these experiments agree with those obtained in pulsed electron beam experiments (Chapters Three and Four).

There are some experimental data which were interpreted in terms of the existence of long-lived "excited water molecules" capable of reacting with even moderate concentrations of scavengers. Sworski [33] considers such a case and concludes that the lifetime of the species is 10^{-9} to 10^{-10} sec. On the basis of our present knowledge of the excited states of the water molecule, it is difficult to explain the existence of such "excited water." It has also been shown that the homogeneous kinetic arguments from which the above conclusion was drawn can be effectively replaced by a nonhomogeneous kinetic treatment applied to the generally accepted reactive species [34], so that there is no need for introducing long-lived excited species into the reaction scheme.

Some other information also may be useful in completing the picture of excited water molecules in the liquid state. Studying the light emission in aqueous solutions of T_2O, Czapski and Katakis [35] concluded that it does not come from deactivation of excited H_2O or OH species. Shaede and Walker [36] have arrived at a similar conclusion by examining luminescence from water irradiated with an extremely intense and very short pulse of high-energy electrons. Analyzing different possibilities for eventual energy transfer from excited water molecules to solutes, Stein [11,12] concludes that aqueous systems are particularly devoid of evidence for such an effect.

Concerning the fate of the H_2O^+ species formed in the primary ionization $H_2O \rightsquigarrow e^- + H_2O^+$, there is no doubt that in such a markedly polar medium as water it should be sought in the reaction

$$H_2O^+ + H_2O \longrightarrow H_3O^+ + OH \qquad (9)$$

Lampe et al. [37] estimate that it takes place in 1.6×10^{-14} sec. As will be seen in the next section, this reaction is even faster than the neutralization reaction ($e^- + H_2O^+$).

We do not know very much about the hydrogen ions formed in reaction (9), but the progress accomplished in understanding the motion of ions [38] promises more knowledge on this subject. The first quantitative data on

$H_3O_{aq}^+$ formation have been reported in papers on hydrogen ion yields in deaerated aqueous *p*-bromphenol solution [39] and the transient conductivity changes in aqueous solutions during and after irradiation with pulsed electron beam [40].

III. Fate of Electrons Produced in the Ionization of Water

An electron detached in the ionization of a water molecule may possess sufficient energy to produce further ionizations and excitations. The energy is dissipated until it becomes lower than the threshold of electronic excitations. Such electrons are often called subexcitation electrons. They cannot transfer any energy to the electronic system of the molecules with which they collide, and their further energy loss must then be mainly by exciting the vibrational and rotational modes of such molecules either directly or via compound negative-ion states. It seems that quite a considerable part of the total energy absorbed from high-energy radiation (10–15%) is dissipated by subexcitation electrons; they ultimately end as thermalized electrons. The fate of subexcitation and thermalized electrons is closely related to the formation of primary radical products in water radiolysis.

Samuel and Magee [41] were the first to consider theoretically the problem in more detail. They calculated that the time required for an electron to return to the parent ion from a distance varying between 5 and 100 Å ranges from 1.25×10^{-15} to 1.23×10^{-11} sec. Assuming a relatively large scattering cross section for electrons in the energy range of interest (a few electron volts), they conclude that the electron cannot go very far without undergoing large scattering deflection, especially as the scattering is predominantly inelastic. The authors estimate that a time of the order of 10^{-13} sec suffices for the return of the electron to the vicinity of the positive ion, where the following reaction takes place:

$$H_2O^+ + e^- \longrightarrow H_2O^* \tag{10}$$

This means that after the recapture of the electron the qualitative difference between the primary ionized and excited water molecules disappears. There are, instead, only water molecules with different amounts of excitation energy, which, in some cases, dissociate as indicated by Eq. (7) to give H and OH radicals. It should be noted that the place of origin and the initial distribution of these two primary species is the same. Magee [42] subsequently introduced some modifications in order to show how the hydrated electron fits into the above recapture hypothesis. He takes reaction (9) to be faster than reaction (10). At the instant of formation of H_3O^+, the dielectric has not yet relaxed, and the H_3O^+ ion is not a normal solvated ion at the moment when the recapture of the electron occurs:

$$H_3O^+ + e^- \longrightarrow H_3O \tag{11}$$

The unstable, partially solvated entity H_3O can undergo two thermal dissociation reactions,

$$H_3O \longrightarrow \begin{cases} H + H_2O \\ e_{aq}^- + H_3O^+ \end{cases} \quad (12)$$

It is also possible that the H_3O remains as such during the intraspur reactions and that the formation of H atoms depends on the steady-state concentration of H_3O,

$$H_3O^+ + e_{aq}^- \rightleftharpoons H_3O \longrightarrow H + H_2O \quad (13)$$

According to Platzman, most of the electrons escape the Coulomb forces of the positive ion parents and become hydrated [2,20,43,44]. Since H_2O^+ reacts according to Eq. (9), the points of formation of the OH radical and of the hydrated electron are well separated, which is basically different from the preceding hypothesis. Mozumder has calculated that the probability of electron escape in water is 0.521 at room temperature. His theory of neutralization of an isolated ion pair in polar media shows that in water the probabilities of neutralization and solvation are about equal [45,46].

In principle, the fate of the thermalized electron might be attachment to a neutral water molecule and formation of the H_2O^- ion or $H + OH^-$. Magee and Burton [47] have considered this last possibility and arrived at the conclusion that this reaction would be exothermic only if it had at its disposal the solvation energy of the OH^- ion produced. However, these authors think that, since a process more rapid than solvation is in question, no solvation energy can be involved, and hence the reaction under consideration cannot take place. In view of the markedly polar character of water molecules, it is often assumed that it is most likely that the electron, after having been thermalized, gives rise by virtue of its electric field to an orientation of water dipoles and thus becomes hydrated. Since the relaxation time of water dipoles is estimated to be about 10^{-11} sec, this should be the time necessary for formation of a hydrated electron.

According to experimental results [48], the hydrated electron can be observed on a picosecond scale, at 20×10^{-12} sec. This is compatible with theoretical results on the dielectric relaxation and electron solvation, which point to 10^{-13} sec as a possible shortest time for the hydration process [49].

IV. Spatial Distribution of Active Species

If a chemical process is induced thermally or photochemically, then the distribution of active species is homogeneous in space. Radiation-chemical processes are essentially different—in this case active species are produced only along the track of the incident particle, and it is only after they have

IV. SPATIAL DISTRIBUTION OF ACTIVE SPECIES 37

diffused throughout the reaction volume (see Chapter Six) that we have conditions compatible with homogeneous kinetics.

The quantities of active species produced along the track of the incident particle and their distribution in space depend on the type and energy of the radiation. In Section I we have seen that as the velocity of the incident particle decreases we may expect the active species to be produced closer together, or, which amounts to the same thing, the higher the velocity, the larger the spacing of the active species. The distribution of events along the track of an ionizing particle may be presented as a "string of beads"—according to a greatly simplified picture, each bead corresponds to the same fixed amount of energy deposited, contains on the average the same number of active species, has the same diameter, and is at the same, fixed distance between adjacent beads. As will be seen in Chapter Six, this simplified approach has been quite successful in the discussion of diffusion-controlled reactions of radicals in water and aqueous solutions.

It is easily understood that this picture of a "string of equidistant beads," a constant spur size, etc., is rather a rough simplification [21a–c]. This is especially so if one takes into account the large energy spectrum of secondary electrons produced along the track of the ionizing particle in water. It is quite evident that along a track of radiation of low linear energy transfer (LET), like ^{60}Co γ-rays, one will find not only isolated spurs, but also regions composed of overlapping spurs produced by low-energy electrons which, in some cases, produce continuous cylindrical regions of activated molecules. The problem is then to determine the contributions of these different types of energy localizations to the total energy deposited. This was considered by Mozumder and Magee [50]. The authors concluded that one must take into consideration not only isolated spurs for which the energy deposited lies between 6 and 100 eV, but also substantially larger regions. They call these "blobs" when the total energy deposited is in the range from 100 to 500 eV, and "short tracks" when the energy transferred lies between 500 and 5000 eV. Because of the higher radical concentration, blobs are expected to behave differently from spurs in reactions in the presence of solute. Short tracks, on the other hand, represent roughly finite cylindrical regions of high LET embedded sporadically in the main track due to an ionizing particle of essentially low LET. Figure 2.3 presents a graphical plot of the percentage of energy deposited in spurs, blobs, and short tracks as a function of electron energy in water [50].

It can be seen that at low primary energies the track as a whole looks like a single short track. In this case, spur and blob contributions are negligible. On the other hand, at high initial particle energies the isolated spur contribution predominates. Blob and short-track contributions are low in this case, but not negligible.

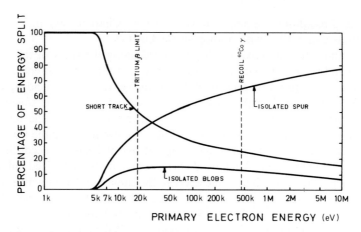

Fig. 2.3. Different types of energy localizations in water as a function of primary electron energy. (After Mozumder and Magee [50].)

It is interesting to note the case of 440 keV electrons, which play the most important role in the energy deposition of ^{60}Co γ-rays in water (see Section IA). In this case, about 64% of the energy is deposited in the form of isolated spurs, about 11% through blobs, and about 25% in the form of short tracks.

It is generally accepted that due to the high ionization densities and spur overlapping, the tracks of heavy particles can be considered to be cylindrical. Mozumder *et al.* [51] have considered in detail the structure of such heavy-particle tracks in water. The track model proposed also consists of a cylinder, but it has a core of high ionization density surrounded by emergent electrons of lower ionization density.

V. Brief Survey of Important Processes Leading to the Formation of Primary Free-Radical Products in Water Radiolysis

The overall process of producing chemical changes by the use of ionizing radiation starts with the bombardment of water or aqueous solution by the radiation and terminates with the reestablishment of chemical equilibrium. This process is usually divided into three stages.

● The physical stage consists of energy transfer to the system. Its duration is of the order of 10^{-15} sec or less. The incident radiation produces, directly or indirectly, ionization of water,

$$H_2O \ \rightsquigarrow \ e^- + H_2O^+ \tag{14}$$

V. FORMATION OF PRIMARY FREE-RADICAL PRODUCTS

as well as excitation,
$$H_2O \leadsto H_2O^* \tag{15}$$

• The physicochemical stage consists of processes which lead to the establishment of thermal equilibrium in the system. Its duration is usually taken to be of the order of 10^{-11} sec or less. Electrons ejected in the ionization process become thermalized and hydrated:

$$e^- \longrightarrow e^-_{therm} \longrightarrow e^-_{aq} \tag{16}$$

The H_2O^+ ions undergo a proton transfer reaction with neighboring water molecules,

$$H_2O^+ + H_2O \longrightarrow H_3O^+ + OH \tag{9}$$

and the H_3O^+ becomes hydrated; hence, the point of formation and the initial spatial distribution of $H_3O^+_{aq}$ and OH are essentially the same and different from those for e^-_{aq}. If the volume in which they arose is considered to be spherical, then the radius for the initial distribution of the hydroxyl radicals and $H_3O^+_{aq}$ is about three times smaller than that of the hydrated electrons.

The dissociation of excited water molecules gives the hydrogen atom and hydroxyl radical as main products,

$$H_2O^* \longrightarrow H + OH \tag{7}$$

It also gives a low yield of hydrated electrons and, eventually, molecular hydrogen. It seems, however, that the contribution of excited water molecules to the formation of primary free-radical products in water radiolysis is of minor importance in comparison with that of the ionization processes.

• The chemical stage consists of diffusion away from the point of origin and chemical reaction of primary species (e^-_{aq}, OH, $H_3O^+_{aq}$, and H), and leads to the establishment of chemical equilibrium. It begins in the spur about 10^{-11} sec after the passage of the radiation and about 10^{-10} sec in the bulk of the solution. Table 2.2 summarizes important chemical reactions of the primary species, which will be considered later in more detail (Chapters Three and Four).

The time scale of the various processes by which the active species redistribute, or get rid of their excess energy, is presented in Fig. 2.4. It may be seen that the transfer of an amount of energy comparable to the electronic binding energy of a molecule takes a time of the same order of magnitude as the electronic oscillation period, that is, 10^{-16} to 10^{-15} sec. Any change in a molecule which involves the displacement of atoms—that is, the stretching of a bond—is limited in speed by the inertia of the atoms and the binding

TABLE 2.2 REACTIONS OF FREE RADICALS IN IRRADIATED WATER

Reaction	Rate constant, M^{-1} sec^{-1}	pH	Reference
$e_{aq}^- + e_{aq}^- \xrightarrow{2H_2O} H_2 + 2OH^-$	5.5×10^9	13.3	[52]
	5×10^9	10–13	[53,54]
	6×10^9	11	[55]
$e_{aq}^- + H \xrightarrow{H_2O} H_2 + OH^-$	2.5×10^{10}	10.5	[52]
	3×10^{10}	10.9	[54]
$e_{aq}^- + OH \longrightarrow OH^-$	3.0×10^{10}	11	[52,54]
$e_{aq}^- + O^- \longrightarrow 2OH^-$	2.2×10^{10}	13	[52]
$e_{aq}^- + H_3O^+ \longrightarrow H + H_2O$	2.06×10^{10}	2.1–4.3	[56]
	2.36×10^{10}	4–5	[53,54]
	2.2×10^{10}	—	[57, 58]
	2.26×10^{10}	4.1–4.7	[59]
$e_{aq}^- + H_2O_2 \longrightarrow OH + OH^-$	1.23×10^{10}	7	[53,54]
	1.36×10^{10}	11	[56,60]
	1.1×10^{10}	—	[58]
	1.3×10^{10}	11	[61]
$e_{aq}^- + HO_2^- \longrightarrow O^- + OH^-$	3.5×10^9	13	[62]
$e_{aq}^- + H_2O \longrightarrow H + OH^-$	16	8.3–9.0	[63]
$H + H \longrightarrow H_2$	1.5×10^{10}	0.2–0.8 N H$_2$SO$_4$	[64]
	1.0×10^{10}	2.1	[65]
	7.75×10^9	3	[66]
	1.3×10^{10}	0.4–3	[67]
	1.25×10^{10}	2–3	[68]
$H + OH \longrightarrow H_2O$	3.2×10^{10}	0.4–3	[67]
$H + OH^- \longrightarrow e_{aq}^-$	1.8×10^7	11.5	[52]
	2.2×10^7	11–13	[69]
$H + H_2O_2 \longrightarrow H_2O + OH$	5×10^7	acid	[70]
	1.6×10^8	0.4–3	[67]
	9×10^7	2	[65]
	4×10^7	—	[71]
$OH + OH \longrightarrow H_2O_2$	6×10^9	0.4–3	[67,72,73]
	4×10^9	7	[74]
	5×10^9	—	[66,75]
$O^- + O^- \xrightarrow{H_2O} HO_2^- + OH^-$	1×10^9	13	[72,73]
$OH + OH^- \longrightarrow O^- + H_2O$	3.6×10^8	—	[70,76]
$OH + H_2O_2 \longrightarrow HO_2 + H_2O$	4.5×10^7	7	[74]
	1.2×10^7	0.4–3	[67]
	2.25×10^7	—	[77]
$O^- + HO_2^- \longrightarrow O_2^- + OH^-$	7×10^8	13	[62]
	2.74×10^8	13	[78]
$OH + H_2 \longrightarrow H + H_2O$	6×10^7	7	[75]
	4.5×10^7	7	[74]
$O^- + H_2 \longrightarrow H + OH^-$	8×10^7	13	[52]
$H_3O^+ + OH^- \longrightarrow 2H_2O$	14.3×10^{10}	—	[79,80]
	15×10^{10}	—	[40]
	4.4×10^{10} [a]	—	[40]

[a] Low value obtained in neutral solution with high-intensity pulses.

V. FORMATION OF PRIMARY FREE-RADICAL PRODUCTS

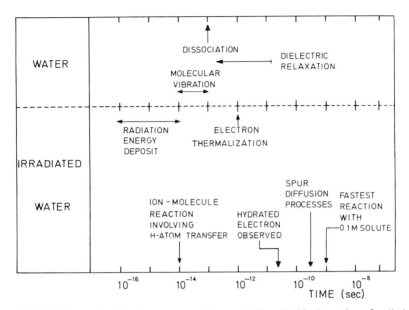

Fig. 2.4. Time scale of various processes that may be involved in the action of radiation on water.

force between them. It must take a time comparable to the normal vibrational period for that bond, that is, 10^{-14} to 10^{-13} sec. The period of rotation of the molecule as a whole is much longer. We are concerned with rotation chiefly in connection with the hydration phenomenon, where water molecules have to orient themselves in the electric field of the electron. For water, this relaxation time is 10^{-11} sec and less at room temperature. If an excited molecule gets rid of its excess energy by emission of radiation, this fluorescence will usually occur some 10^{-9} to 10^{-8} sec after the excitation. If the molecule dissociates, this will take a time at least as long as the vibrational half-period and, if the molecule has many degrees of freedom, it may take a much longer one. The dissociation time of the water molecule has been estimated to be about 10^{-13} sec, although a value of 10^{-14} sec is frequently found in the literature. For the sake of comparison, the life of a pseudo-first-order radical–solute reaction is given at 0.1 M scavenger concentration.

The expression for radiation-induced water decomposition,

$$H_2O \;\leadsto\; H_3O^+_{aq}, \;\; OH, \;\; e^-_{aq}, \;\; H, \;\; H_2O_2, \;\; H_2 \tag{17}$$

represents the state in irradiated water about 10^{-9} sec after the passage of high-energy radiation, when the reactions in the spurs, blobs, and short tracks are practically terminated. These products are found in irradiated water irrespective of the type and energy of radiation. Their amounts per 100 eV absorbed, the so-called primary product yields ($G_{H_3O^+}$, G_H, G_{OH},

$G_{e_{aq}^-}$, G_{H_2}, and $G_{H_2O_2}$) depend, however, on the LET of radiation and on other parameters that will be discussed in Chapter Five.

The survey of primary processes given here presents a working hypothesis which has furnished satisfactory explanations for most experimental observations. However, some facts still remain unexplained and other concepts have been proposed [*81–83*].

References

1. H. A. Bethe and J. Ashkin, Passage of Radiations through Matter. *In* "Experimental Nuclear Physics" (E. Segré ed.), Vol. 1, p. 166. Wiley, New York, 1953.
2. R. L. Platzman, Energy Spectrum of Primary Activations in the Action of Ionizing Radiation. *In* "Radiation Research 1966" (G. Silini, ed.), p. 20. North-Holland Publ. Amsterdam, 1967.
3. J. Kistemaker, F. J. De Heer, J. Sanders, and C. Snoek, Energy Deposition in Matter by Slow Heavy Particles. *In* "Radiation Research 1966" (G. Silini, ed.), p. 68. North-Holland Publ., Amsterdam, 1967.
4. A. Ore, The Role of Multiple Ionization in Radiation Action. In "Radiation Research 1966" (G. Silini, ed.), p. 54. North–Holland Publ., Amsterdam, 1967.
5. A. Mozumder, Charged Particle Tracks and Their Structure. *Advan. Radiat. Chem.* **1**, 1 (1969).
6. W. P. Helman, A. Mozumder, and A. Ross, eds., "Proceedings of the Symposium on Very Early Effects." AECD No. COO–38–738, Buenos Aires, 1970.
7. A. H. Breger, ed., "Osnovy Radiatsionno-khimicheskogo apparato-stroenia," p. 239. Atomizdat, Moskva, 1967.
8. R. H. Schuler and A. O. Allen, Radiation Chemistry Studies with Cyclotron Beams of Variable Energy: Yields in Aerated Ferrous Sulfate Solution. *J. Amer. Chem. Soc.* **79**, 1565 (1957).
9. J. Weiss and W. Bernstein, Energy Required to Produce One Ion Pair for Several Gases. *Phys. Rev.* **98**, 1828 (1955).
10. A. O. Allen, "The Radiation Chemistry of Water and Aqueous Solutions," p. 3. Van Nostrand-Reinhold, Princeton, New Jersey, 1961.
11. G. Stein, Excitation and Ionization. Some Correlations between the Photo and Radiation Chemistry of Liquids. *In* "The Chemistry of Ionization and Excitation" (G. R. A. Johnson and G. Scholes, eds.), p. 25. Taylor and Francis, London, 1967.
12. G. Stein, ed. Excitation and Ionization by Ionizing Radiations in Water and Aqueous Solutions. *In* "Radiation Chemistry of Aqueous Systems," p. 83. Wiley (Interscience), New York, 1968.
13. D. Lewis and W. H. Hamill, Evidence for the Triplet State of Water by Electron-Reflection Spectroscopy. *J. Chem. Phys.* **51**, 456 (1969).
14a. F. Fiquet-Fayard, Radiolyse de l'eau légère et de l'eau tritiée par les rayons γ en présence de méthacrylate de méthyle. I. Discussion des réactions élémentaires de la dissociation de l'eau excitée et ionisée. *J. Chim. Phys.* **57**, 453 (1960).
14b. F. Fiquet-Fayard, Study of the Water Molecule in the Gaseous Phase, Excited States of H_2O, H_2O^+ and H_2O^-. *In* "Radiation Chemistry of Aqueous Systems" (G. Stein, ed.), p. 9. Wiley (Interscience), New York, 1968.
15a. I. Santar and J. Bednář, Theory of Radiation Chemical Yield. I. Radiolysis of Water Vapour. *Collect. Czech. Chem. Commun.* **32**, 953 (1967).
15b. I. Santar and J. Bednář, Theory of the Radiation Chemical Yield. II. Derivation of the Fundamental Relations. *Collect. Czech. Chem. Commun.* **33**, 1 (1968).

15c. I. Santar and J. Bednář, Theory of the Radiation Chemical Yield. III. The Effect of Molecular Nature on Primary Yields. *Collect. Czech. Chem. Commun.* **34**, 1 (1969).
16. A. Johannin-Gilles and B. Vodar, Sur le spectre d'absorption de la vapeur d'eau dans l'ultraviolet de Schumann. *J. Phys. Radium* **15**, 223 (1954).
17. K. Watanabe, Ionization Potentials of Some Molecules. *J. Chem. Phys.* **26**, 542 (1957).
18. C. Wingate, W. Gross, and G. Failla, Relative Beta Ray Energy Loss per Ion Pair Produced in Water Vapor and in Air. *Radiat. Res.* **8**, 411 (1958).
19. R. L. Platzman, Superexcited States of Molecules. *Radiat. Res.* **17**, 419 (1962).
20. R. L. Platzman, Superexcited States of Molecules, and the Primary Action of Ionizing Radiation. *Vortex* **23**, 372 (1962).
21a. I. Santar and J. Bednář, The Optical Approximation, Primary Radiation Chemical Yields, and Structure of the Track of an Ionizing Particle. *Advan. Chem. Ser.* **81**, 523 (1968).
21b. I. Santar and J. Bednář, Theory of Radiation Chemical Yield. V. Initial Structure of the Track of a Fast Electron in a Dense Medium. *Int. J. Radiat. Phys. Chem.* **1**, 133 (1969).
21c. I. Santar and J. Bednář, Effect of Chemical Composition and of Initial Energy on the Physical Track Structure for Fast Electrons. *Ch. Part. Tr. Solid Liq.* 62 (1970).
22. C. E. Melton, Radiolysis of Water Vapor in a Wide Range Radiolysis Source of a Mass Spectrometer. I. Individual and Total Cross Sections for the Production of Positive Ions, Negative Ions, and Free Radicals by Electrons. *J. Phys. Chem.* **74**, 582 (1970).
23. R. E. Verrall and W. A. Senior, Vacuum-Ultraviolet Study of Liquid H_2O and D_2O. *J. Chem. Phys.* **50**, 2746 (1969).
24. L. R. Painter, R. D. Birkhoff, and E. T. Arakawa, Optical Measurements of Liquid Water in the Vacuum Ultraviolet. *J. Chem. Phys.* **51**, 243 (1969)..
25. J. Barret and J. H. Baxendale, The Photolysis of Liquid Water. *Trans. Faraday Soc.* **56**, 37 (1960).
26. U. Sokolov and G. Stein, Photolysis of Liquid Water at 1470 Å. *J. Chem. Phys.* **44**, 2189 (1966).
27. U. Sokolov and G. Stein, Photolysis of Liquid Water at 1849 Å. *J. Chem. Phys.* **44**, 3329 (1966).
28. M. Cottin, J. Masanet, and C. Vermeil, Photolyse de la vapeur d'eau pure et en présence de deutérium aux longueurs d'onde de 1236 Å et 1470 Å. *J. Chim. Phys.* **63**, 959 (1966).
29. J. R. McNesby, I. Tanaka, and H. Okabe, Vacuum Ultraviolet Photochemistry. III. Primary Processes in the Vacuum Ultraviolet Photolysis of Water and Ammonia. *J. Chem. Phys.* **36**, 605 (1962).
30. J. Masanet and C. Vermeil, Photolyse de la vapeur d'eau légère et deutériée dans l'ultraviolet lointain. II. Réactions des atomes d'hydrogène et de deutérium "chaud." *J. Chim. Phys.* **66**, 1249 (1969).
31. N. Getoff and G. O. Schenck, Primary Products of Liquid Water Photolysis at 1236, 1470 and 1849 Å. *Photochem. Photobiol.* **8**, 167 (1968).
32. J. W. Boyle, J. A. Ghormley, C. J. Hochanadel, and J. F. Riley, Production of Hydrated Electrons by Flash Photolysis of Liquid Water with Light in the First Continuum. *J. Phys. Chem.* **73**, 2886 (1969).
33. T. J. Sworski, Kinetic Evidence that "Excited Water" is Precursor of Intraspur H_2 in the Radiolysis of Water. *Advan. Chem. Ser.* **50**, 263 (1965).
34. Z. D. Draganić and I. G. Draganić, On the Origin of Primary Hydrogen Peroxide Yield in the γ Radiolysis of Water. *J. Phys. Chem.* **73**, 2571 (1969).

35. G. Czapski and D. Katakis, Light Emission from Aqueous Solutions of T_2O. *J. Phys. Chem.* **70**, 637 (1966).
36. E. A. Shaede and D. C. Walker, Cerenkov Radiation From the Pulse Radiolysis of Water. *Int. J. Radiat. Phys. Chem.* **1**, 307 (1969).
37. F. W. Lampe, F. H. Field, and J. L. Franklin, Reaction of Gaseous Ions. IV. Water. *J. Amer. Chem. Soc.* **79**, 6132 (1957).
38. L. Onsager, The Motion of Ions: Principles and Concepts. *Science* **166**, 1359 (1969).
39. B. Čerček and M. Kongshaug, Hydrogen Ion Yields in the Radiolysis of Neutral Aqueous Solutions. *J. Phys. Chem.* **73**, 2056 (1969).
40. K. H. Schmidt and S. M. Ander, Formation and Recombination of H_3O^+ and Hydroxide in Irradiated Water. *J. Phys. Chem.* **73**, 2846 (1969).
41. A. H. Samuel and J. L. Magee, Theory of Radiation Chemistry. II. Track Effects in Radiolysis of Water. *J. Chem. Phys.* **21**, 1080 (1953).
42. J. L. Magee, Discussion. *Radiat. Res. Suppl.* **4**, 20 (1964).
43. R. L. Platzman, Energy Transfer from Secondary Electrons to Matter. *In* "Basic Mechanisms in Radiobiology. II. Physical and Chemical Aspects" (J. L. Magee, N. D. Kamen, and R. L. Platzman, eds.), p. 22. National Research Council, Publ. No. 305, Washington, D.C., 1953.
44. E. J. Hart and R. J. Platzman, Radiation Chemistry. *In* "Mechanisms of Radiobiology" (M. Errera and A. Forssberg, eds.), Vol. 1, p. 93. Academic Press, New York, 1961.
45. A. Mozumder, Neutralization of Isolated Ion Pair in Polar Media. I. Theory for the Yield of Solvated Electrons. *J. Chem. Phys.* **50**, 3153 (1969).
46. A. Mozumder, Neutralization of Isolated Ion Pair in Polar Media. II. Evolution of the Neutralization Process. *J. Chem. Phys.* **50**, 3162 (1969).
47. J. L. Magee and M. Burton, Elementary Processes in Radiation Chemistry. II. Negative Ion Formation by Electron Capture in Neutral Molecules. *J. Amer. Chem. Soc.* **73**, 523 (1951).
48. M. J. Bronskill, R. K. Wolff, and J. W. Hunt, Subnanosecond Observations of the Solvated Electron. *J. Phys. Chem.* **73**, 1175 (1969).
49. A. Mozumder [6], p. 84.
50. A. Mozumder and J. L. Magee, Model of Tracks of Ionizing Radiations for Radical Reaction Mechanisms, *Radiat. Res.* **28**, 203 (1966).
51. A. Mozumder, A. Chatterjee, and J. L. Magee, Theory of Radiation Chemistry, IX. Model and Structure of Heavy Particle Tracks in Water. *Advan. Chem. Ser.* **81**, 27 (1968).
52. M. S. Matheson and J. Rabani, Pulse Radiolysis of Aqueous Hydrogen Solutions. I. Rate Constants for Reaction of e_{aq}^- with Itself and Other Transients. II. The Interconvertibility of e_{aq}^- and H. *J. Phys. Chem.* **69**, 1324 (1965).
53. S. Gordon, E. J. Hart, M. S. Matheson, J. Rabani, and J. K. Thomas, Reaction Constants of the Hydrated Electron. *J. Amer. Chem. Soc.* **85**, 1375 (1963).
54. S. Gordon, E. J. Hart, M. S. Matheson, J. Rabani, and J. K. Thomas, Reactions of the Hydrated Electron. *Discuss. Faraday Soc.* **36**, 193 (1963).
55. K. Schmidt and E. J. Hart, A Compact Apparatus for Photogeneration of Hydrated Electrons. *Advan. Chem. Ser.* **81**, 267 (1968).
56. J. P. Keene, The Absorption Spectrum and Some Reaction Constants of the Hydrated Electron. *Radiat. Res.* **22**, 1 (1964).
57. J. H. Baxendale, E. M. Fielden, and J. P. Keene, The Pulse Radiolysis of Aqueous Solutions of Some Inorganic Compounds, *Proc. Roy. Soc. (London)* **A286**, 320 (1965).
58. B. Čerček, Activation Energies for Reactions of the Hydrated Electron. *Nature* **223**, 491 (1969).

REFERENCES

59. L. M. Dorfman and I. A. Taub, Pulse Radiolysis Studies. III. Elementary Reactions in Aqueous Ethanol Solution. *J. Amer. Chem. Soc.* **85**, 2370 (1963).
60. B. Hickel and K. H. Schmidt, Kinetic Studies with Photogenerated Hydrated Electrons in Aqueous Systems Containing Nitrous Oxide, Hydrogen Peroxide, Methanol or Ethanol. *J. Phys. Chem.* **74**, 2470 (1970).
61. E. J. Hart and E. M. Fielden, Submicromolar Analysis of Hydrated Electron Scavengers. *Advan. Chem. Ser.* **50**, 253 (1965).
62. W. D. Felix, B. L. Gall, and L. M. Dorfman, Pulse Radiolysis Studies. IX. Reactions of the Ozonide Ion in Aqueous Solution. *J. Phys. Chem.* **71**, 384 (1967).
63. E. J. Hart, S. Gordon, and E. M. Fielden, Reaction of the Hydrated Electron with Water. *J. Phys. Chem.* **70**, 150 (1966).
64. H. A. Schwarz, Absolute Rate Constants for Some Hydrogen Atom Reactions in Aqueous Solution. *J. Phys. Chem.* **67**, 2827 (1963).
65. J. P. Sweet and J. K. Thomas, Absolute Rate Constants for H Atom Reactions in Aqueous Solutions. *J. Phys. Chem.* **68**, 1363 (1964).
66. P. Pagsberg, H. Christensen, J. Rabani, G. Nilsson, J. Fenger, and S. O. Nielsen, Far-Ultraviolet Spectra of Hydrogen and Hydroxyl Radicals from Pulse Radiolysis of Aqueous Solutions. Direct Measurement of the Rate of H + H. *J. Phys. Chem.* **73**, 1029 (1969).
67. H. Fricke and J. K. Thomas, Pulsed Electron Beam Kinetics. *Radiat. Res. Suppl.* **4**, 35 (1964).
68. J. Rabani and D. Meyerstein, Pulse Radiolytic Studies of the Competition H + H and H + Ferricyanide. The Absolute Rate Constants. *J. Phys. Chem.* **72**, 1599 (1968).
69. J. Rabani, The Interconvertibility of e_{aq}^- and H. *Advan. Chem. Ser.* **50**, 242 (1965).
70. C. J. Hochanadel, Photolysis of Dilute Hydrogen Peroxide Solution in the Presence of Dissolved Hydrogen and Oxygen. *Radiat. Res.* **17**, 286 (1962).
71. J. K. Thomas, The Rate Constants for H Atom Reactions in Aqueous Solutions. *J. Phys. Chem.* **67**, 2593 (1963).
72. J. Rabani and M. S. Matheson, The Pulse Radiolysis of Aqueous Solutions of Potassium Ferrocyanide. *J. Phys. Chem.* **70**, 761 (1966).
73. J. Rabani and M. S. Matheson, Pulse Radiolytic Determination of pK for Hydroxyl Ionic Dissociation in Water. *J. Amer. Chem. Soc.* **86**, 3175 (1964).
74. H. A. Schwarz, A Determination of some Rate Constants for the Radical Processes in the Radiation Chemistry of Water. *J. Phys. Chem.* **66**, 255 (1962).
75. J. K. Thomas, J. Rabani, M. S. Matheson, E. J. Hart, and S. Gordon, Absorption Spectrum of the Hydroxyl Radical. *J. Phys. Chem.* **70**, 2409 (1966).
76. F. S. Dainton and S. A. Sills, The Rates of Some Reactions of Hydrogen Atoms in Water at 25°C. *Proc. Chem. Soc. (London)* 223 (1962).
77. J. K. Thomas, Rates of Reaction of the Hydroxyl Radical. *Trans. Faraday Soc.* **61**, 702 (1965).
78. J. Rabani, Pulse Radiolysis of Alkaline Solutions. *Advan. Chem. Ser.* **81**, 131 (1968).
79. M. Eigen and L. DeMaeyer, Untersuchungen über die Kinetik der Neutralisation. I. *Z. Elektrochem.* **59**, 986 (1955).
80. G. Ertl and H. Gerischer, Ein Vergleich der Kinetik der Neutralisationreaktionen des leichten und schweren Wassers. *Z. Elektrochem.* **66**, 560 (1962).
81. M. Anbar, Water and Aqueous Solutions. *In* "Fundamental Processes in Radiation Chemistry" (P. Ausloos, ed.), p. 651. Wiley (Interscience), New York, 1968.
82. J. C. Russell and G. R. Freeman, Yield of Solvated Electrons in the γ Radiolysis of Water + 10% Ethanol: Nonhomogeneous Kinetics of Electron Scavenging in Water. *J. Chem. Phys.* **48**, 90 (1968).
83. W. H. Hamill, A Model for the Radiolysis of Water. *J. Phys. Chem.* **73**, 1341 (1969).

In the preceding chapter we have seen that very soon after the passage of radiation, two short-lived reducing species are present in irradiated water: the hydrated electron and the hydrogen atom. We shall try to present here in more detail what is known about their nature and behavior in radiation-chemical processes in aqueous solutions.

As was seen in the concise review of the development of the radiation chemistry of water (Chapter One, Section III), molecular and atomic hydrogen were assumed to be primary products. The Samuel–Magee model of water radiolysis [1, 2], which met the needs of radiation chemists over a long period, took into consideration only the existence of H atoms as short-lived reducing species, ruling out the possibility that the secondary electron could escape from the field of the parent ion. It is true that even at that time it was suggested that a thermalized electron may move sufficiently far away from the parent ion to become hydrated [3, 4]. However, it was only in the 1960s that chemical experiments were carried out [5–9] which called attention to the fact that there are two kinds of short-lived reducing species differing from each other both in reaction rates and in the mode in which they react. It was soon shown that one of the species has a negative charge [10, 11]. When it was established that it also has a characteristic absorption spectrum in the visible region [12–14], new possibilities were opened up for the study of very fast reactions.

In connection with the subject dealt with in this chapter, we note that the amount of literature on pulse radiolysis and on the hydrated electron is particularly vast. We refer the reader interested in more detail to the material from two symposia [15, 16] and the complete bibliography in this field up to 1969 [17]. One should also mention the review articles on the hydrated electron [18–26].

The question of the origin of atomic and molecular hydrogen, i.e., of the way in which these are produced in the irradiation of water, is still a matter of controversy, particularly as regards a possible role of excited molecules. In this connection we shall present various experimental data, although these often obscure rather than clarify the picture of the primary act of water radiolysis.

CHAPTER THREE

Primary Products of Water Radiolysis: Short-Lived Reducing Species— the Hydrated Electron, the Hydrogen Atom, and Molecular Hydrogen

I. Properties of the Hydrated Electron

A. *The Simplest Elementary Negative Ion in Aqueous Solutions*

In applying the Brønsted–Bjerrum theory of ionic reactions to the behavior of the hydrated electron it is assumed that it has a fully developed ionic atmosphere before it reacts; numerous experiments confirm that the hydrated electron actually behaves like a normal univalent ion. It should be noted, however, that in some cases the electron may react with the solute before it has completed the formation of its ionic atmosphere [27]: This is the case when its reactivity toward the solute is very high and its half-life before reaction is very short. Relation (1) defines the dependence of its half-life on the reactivity,

$$t_{1/2} = \frac{0.693}{\sum k_i C_i} \text{ sec} \tag{1}$$

and Eq. (2) gives the time of relaxation of the ionic atmosphere,

$$\tau = 3.55 \times 10^{-9} \frac{\sum Z_i}{\mu \sum \lambda_i} \text{ sec} \tag{2}$$

Here, Z_i is the charge on the ionic species i, λ_i is the ionic mobility, k_i is the rate constant for the reaction between the ith species and the electron, C_i is the ionic concentration, $\mu = 0.5 \sum C_i Z_i^2$. The summations are over all the ionic species present. It is obvious that if the rate constant is large enough, then a concentration might be reached at which τ is greater than $t_{1/2}$.

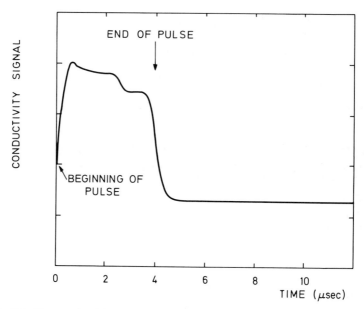

Fig. 3.1. The conductivity of a dilute aqueous solution [Ba(OH)$_2$, 4×10^{-5} N] during and after irradiation with a pulse of high-energy electrons. (After Schmidt and Buck [28].)

It has been shown [28] that the conductivity of carefully purified water rises at the beginning of the pulse of radiation and diminishes after it. This is illustrated in Fig. 3.1, where the experimental data are given for measurements in a 4×10^{-5} N solution of Ba(OH)$_2$. The measurements were carried out during and after irradiation with an electron pulse lasting about 4 μsec at a mean dose rate of 4×10^{22} eV ml^{-1} sec^{-1}. The absolute ionic mobility of the hydrated electron produced by the radiation was 1.8×10^{-3} cm^2 V^{-1} sec^{-1}. The diffusion constant is calculated from this to be

$$D_{e_{aq}^-} = 4.75 \times 10^{-5} \quad \text{cm}^2 \text{ sec}^{-1} \; (\pm 10\%)$$

The redox potential for the hydrated electron has been calculated by using the following cycle [29]:

$$e_{aq}^- + H_{aq}^+ \longrightarrow \tfrac{1}{2} H_2 \qquad \Delta F_3 = ? \qquad (3)$$

$$H \longrightarrow \tfrac{1}{2} H_2 \qquad \Delta F_4 = -48.5 \text{ kcal mole}^{-1} \qquad (4)$$

$$H^+ + OH^- \longrightarrow H_2O \qquad \Delta F_5 = -19.3 \text{ kcal mole}^{-1} \qquad (5)$$

$$e_{aq}^- + H_2O \underset{k_{6b}}{\overset{k_{6f}}{\rightleftarrows}} H + OH^- \qquad (6)$$

The change in free energy in this last case can be calculated, since it is known

I. PROPERTIES OF THE HYDRATED ELECTRON

that $k_{6f} = 16 \, M^{-1} \, \text{sec}^{-1}$ and $k_{6b} = 2.2 \times 10^7 \, M^{-1} \, \text{sec}^{-1}$, that is, $k_{6f}/k_{6b} = K = 7.27 \times 10^{-7}$

$$\Delta F_6 = 8.4 \text{ kcal mole}^{-1}$$

Summing up (4)–(6), we obtain the change in free energy for reaction (3):

$$\Delta F_3 = \Delta F_4 + \Delta F_5 + \Delta F_6 = -48.5 \text{ kcal mole}^{-1} - 19.3 \text{ kcal mole}^{-1} + 8.4 \text{ kcal mole}^{-1} = -59.4 \text{ kcal mole}^{-1}$$

Hence the redox potential of the hydrated electron on the conventional scale is calculated to be $E^0 = 2.6$ V, compared with 2.1 V for the H atom. As the data on reaction rate constants also confirm, the hydrated electron is a more powerful reducing agent than the H atom.

Using the above data and the value -260.5 kcal mole^{-1} for the absolute hydration energy of the hydrogen ion, the hydration energy of the electron can be calculated to be -40 kcal mole^{-1} [29] from the following cycle:

$$\begin{array}{ccc} e^-_{aq} + H^+_{aq} & \longrightarrow & (\tfrac{1}{2}H_2)_{aq} \\ \downarrow \, \downarrow & & \searrow \\ e^-_g + H^+_g & \longrightarrow \; H_g \; \longrightarrow & (\tfrac{1}{2}H_2)_g \end{array}$$

This is a surprisingly small hydration energy, and it implies considerable dispersal of the charge of the electron over neighboring molecules. Thermodynamic considerations [30,31] lead to different useful parameters of the hydrated electron; one is the value 2.98 Å [30] as the radius for its charge dispersion.

That the radius of e^-_{aq} is about 3 Å can be seen from a completely different approach which is based on the Debye equation. According to Debye, the rate constant for the reaction between species A and B is

$$k_{AB} = \frac{4\pi r_{AB} D_{AB} N}{1000} \left\{ \left(\frac{Z_A Z_B e^2}{r_{AB} \varepsilon k T}\right) \Big/ \left(\exp\left[\frac{Z_A Z_B e^2}{r_{AB} \varepsilon k T}\right] - 1 \right) \right\}$$

The term preceding the brackets is the encounter rate for uncharged species A and B whose total diffusion constant is D_{AB} and whose encounter radius is r_{AB}; N is Avogadro's number and k_{AB} is the reaction rate constant in liters per mole per second. The term in brackets is the Debye correction to the encounter rate if A and B have charges $Z_A e$ and $Z_B e$, the force between them is coulombic, and the effective dielectric constant of the solution is ε; T is the absolute temperature, k is Boltzmann's constant, and e is the charge on the electron. The equation applied to the case of the reaction $e^-_{aq} + e^-_{aq}$ has r_{AB} as unknown parameter. It can be calculated from experimental data that r_{AB} is just over 6 Å, corresponding to about 3 Å for the charge density radius.

In water, as in other polar solvents, excess electrons are localized within a solvent cavity. The radius of this cavity is different from the charge density radius considered above. It has been calculated [32] to be 1.5 Å for the hydrated electron in water.

B. Absorption Spectrum

An electron slowed down in a liquid of high dielectric constant, the molecules of which cannot easily capture it, will, after thermalization, most likely give rise to an orientation of dipoles by virtue of its electric field. To put it figuratively, when the electron has been slowed down to thermal energy, it is to be expected that the liquid molecules surrounding the electron will tend to align themselves in the electric field of the latter, as shown in Fig. 3.2. Thermal motion will prevent persistent polarization of the molecules adjacent to the electron, but the polarization further on, although weaker, will be more

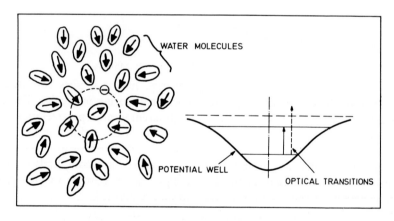

Fig. 3.2. The hydrated electron: a schematic picture of polarization (left) and energy levels (right). (From Boag [22]. Reproduced by permission of North-Holland Publishing Company.)

extensive and more persistent. If we imagine the electron itself to be instantaneously removed from its central position in this assemblage but the molecular polarization to be "frozen in," we can see that a positive potential must exist. In this well, which it has dug for itself, the negative electron is trapped almost as effectively as if it were bound to a positively charged nucleus. A quasi entity is formed in which the electron is bound not to one but to an assemblage of many water molecules. According to quantum theory, the electron can occupy only certain prescribed orbits in this field, i.e., it has

I. PROPERTIES OF THE HYDRATED ELECTRON

quantized energy levels. Transitions between these energy levels give rise to the optical absorption spectrum (Fig. 3.3).

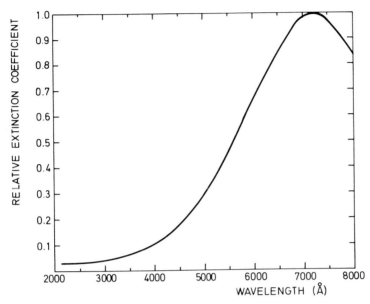

Fig. 3.3. Absorption spectrum of the hydrated electron in water at room temperature. The relative extinction coefficient is normalized to 1.00 at 7200 Å. (After Boyle et al. [37] and Fielden and Hart [34].)

Hart and Boag [12] were the first to identify the absorption spectrum of the solvated electron. They have shown that its form is similar to that of the spectrum possessed by a solvated electron in ammonia. The intensity of absorption does not change if the water contains substances which are known to be unreactive toward reducing primary products of water radiolysis, but it decreases with increasing concentration of substances which are known to effectively react with the latter. Keene has shown that a photocell and an oscilloscope may be of use for observing changes in absorption as a function of time [14].

The absolute value of the molar extinction coefficient of the hydrated electron has been determined in an astute fashion [33], making use of the reaction with tetranitromethane, $e_{aq}^- + C(NO_2)_4 \rightarrow C(NO_2)_3^- + NO_2$, and of the known value of the molar extinction coefficient for the nitroform anion produced in this reaction, $\varepsilon_{3660} = 10,200\ M^{-1}\ cm^{-1}$ at 24°C. The increase in absorption at 3660 Å due to the production of nitroform and the decrease in absorption of hydrated electron at 5780 Å were simultaneously measured after the electron pulse input. From these data the unknown molar extinction coefficient for the hydrated electron was calculated to be $\varepsilon_{5780} = 10{,}600$ $M^{-1}\,cm^{-1}$. It increases with increasing wavelength, with a maximum at about 7200 Å and $\varepsilon_{7000} = 18{,}500\ M^{-1}\,cm^{-1}$ [34]. A new absorption band with a maximum below 2000 Å has been reported [35]. It is worth noticing that ultrafast measurements, with a resolving time of 20 psec (2×10^{-11} sec), confirm the absorption spectrum (measurements between 4000 and 7000 Å) obtained in microsecond times [36]. Furthermore, the absorption spectrum of e_{aq}^- produced in flash-photolytic experiments is identical with the spectrum registered in pulse radiolysis [37].

C. Reactivity

The hydrated electron is a very reactive species, as can be seen from the tables of rate constants. Very often these are close to the diffusion-controlled values. Taking this as well as Eq. (1) into account, it is easily seen that the half-life of the electron, even in dilute solutions, must be very short. In early experiments it was found to be of the order of magnitude of a few tens of microseconds. However, it was obvious that the value observed depends on the presence of impurities in the solution. Molecular oxygen is certainly the most troublesome of these, both because it is difficult to remove from solution and because it is a very efficient radical scavenger. This applies to H_2O_2, which is produced together with the hydrated electron in the process of radiolysis, as well as to the H_3O^+ ion, which arises from the dissociation of water. The concentration of e_{aq}^- also has an effect on its lifetime: The higher the concentration, the shorter the lifetime, because the reaction of recombination is favored. In this connection, it is interesting to consider the experimental conditions and the results relative to the measurement of the rate constant for the reaction [38]

$$e_{aq}^- + H_2O \longrightarrow H + OH^- \qquad [6]$$

The concentration of impurities present was lower than 5×10^{-9} M. The energy input was 1–2 rad per pulse, producing a concentration of only $5 \times 10^{-9}\ M\ e_{aq}^-$.

These measurements established that

$$k_{e_{aq}^- + H_2O} = 16\quad M^{-1}\,sec^{-1}$$

The half-life of the hydrated electron under these conditions is about 800 μsec. As the pH decreases, so does $t_{1/2}$, attaining a value of 300 μsec in neutral water, where $[H_3O^+]$ is about 10^{-7} M. The half-life increases with increasing pH and, depending on other conditions, rises to several tens of

I. PROPERTIES OF THE HYDRATED ELECTRON

milliseconds. All in all, the low reactivity of e_{aq}^- toward water is striking, particularly when the rate constants for its reactions with other substances are borne in mind. The corresponding reaction in heavy water is one order of magnitude slower [39].

All hydrated electron reactions are by definition electron-transfer reactions. In all cases, the primary product of the electron-transfer reaction acquires an additional electron,

$$e_{aq}^- + A^n \longrightarrow A^{n-1} \tag{7}$$

Here, A is an atom or a polyatomic molecule and n is a positive or negative integer or zero. In numerous cases, the primary product is thermodynamically unstable and undergoes further reactions such as protonation, dissociation, disproportionation, or charge transfer. Many hundreds of reactions of the hydrated electron are at present well established. The reaction $e_{aq}^- + O_2 \to O_2^-$ is an example of nondissociative electron capture by a molecule, whereas the reaction $Cu^{2+} + e_{aq}^- \to Cu^+$ illustrates nondissociative electron capture by an ion. Another characteristic type of reaction is dissociative capture, as in the case of $e_{aq}^- + H_2PO_4^- \to H + HPO_4^{2-}$. Tables 3.1 and 3.2 summarize the rate constants for reactions which are most often used in laboratory practice. For detailed information the compilation made by Anbar and Neta [40] may be consulated.

Analysis of the data on rate constants allows some general comments:

- Positive inorganic ions are more reactive than negative ones, and the higher the charge, the higher the rate constant.
- Organic acids react most rapidly when they are not dissociated; the higher the degree of dissociation, the lower the rate constant for reaction with the hydrated electron (e.g., oxalic, formic, and acetic acids).
- Reactive groups in aliphatic compounds are C—Cl, C=O, C=S, and S—S.
- Aromatic compounds (benzene and phenol) react slowly, but some of their derivatives (such as nitrobenzene, benzonitrile, and benzoate ion) may react rapidly.
- Some compounds of interest for radiobiologists (purine, cystamine and thymine) are very reactive, but some react slowly (amino acids).
- Derivatives of aromatic compounds, display a higher reactivity as the character of the ring becomes more positive (i.e., the more pronounced the electrophilic properties of the ring).

The temperature dependence of reactions of the hydrated electron is slight, only a few tenths of a percent per degree (Chapter Five, Section VIII).

TABLE 3.1
RATE CONSTANTS FOR e_{aq}^- REACTIONS WITH SOME INORGANIC SOLUTES AND FREE RADICALS

Reactant	Rate constant, M^{-1} sec^{-1}	pH	Reference
Ag^+	3.5×10^{10}	7	[41]
Al^{3+}	2.0×10^9	6.8	[42]
BrO^-	2.3×10^{10}	13	[43]
BrO_2^-	1.8×10^{10}	13	[43]
BrO_3^-	4.1×10^9	13	[43]
	7.8×10^9	—	[44]
CO	1×10^9	—	[45]
CO_2	7.7×10^9	7	[46]
CO_3^{2-}	$< 10^6$	>9	[47]
CNS^-	$< 10^6$	7	[47]
Cd^{2+}	6.4×10^{10}	4.4–4.5	[48]
ClO^-	7.0×10^9	10.2	[49]
ClO_3^-	2.2×10^8	—	[44]
ClO_4^-	$< 10^6$	10	[47]
Co^{2+}	1.2×10^{10}	—	[50]
	9.5×10^9	4	[48]
Cu^{2+}	3.4×10^{10}	1.5–4.5	[51]
	4.5×10^{10}	3–4	[48]
	3.0×10^{10}	7	[42]
	5.8×10^9	14	[42]
e_{aq}^-	5.5×10^9	13.3	[52]
	5×10^9	10–13	[46, 53]
	6×10^9	11	[54]
	6.5×10^9	13	[55]
Fe^{2+}	3.5×10^8	—	[50]
$Fe(CN)_6^{4-}$	$< 10^5$	—	[56]
$Fe(CN)_6^{3-}$	3×10^9	7, 10.3	[46, 53]
H	2.5×10^{10}	10.5	[52]
	3×10^{10}	10.9	[46]
HCO_3^-	$< 10^6$	—	[47]
H_2O	1.6×10^1	8.3–9	[38]
H_2O_2	1.23×10^{10}	7	[46, 53]
	1.36×10^{10}	11	[57, 58]
	1.1×10^{10}	—	[44]
	1.3×10^{10}	7	[59]

TABLE 3.1—Continued

Reactant	Rate constant, M^{-1} sec^{-1}	pH	Reference
HO_2^-	3.5×10^9	13	[60]
H_3O^+	2.06×10^{10}	2.1–4.3	[57]
	2.36×10^{10}	4–4.6	[46, 53]
	2.2×10^{10}	—	[44, 61]
	2.26×10^{10}	4.1–4.7	[55]
$H_2PO_2^-$	1.1×10^5	6.8	[49]
$H_2PO_3^-$	7.2×10^6	6.7	[49]
$H_2PO_4^-$	7.7×10^6	7.1	[49]
H_2S	1.35×10^{10}	5.5–6	[62]
I_2	5.1×10^{10}	7	[47]
In^{3+}	5.6×10^{10}	1	[63]
K^+	$< 5 \times 10^5$	—	[50]
La^{3+}	3.4×10^8	6.98	[47]
Mn^{2+}	7.7×10^7	—	[50]
MnO_4^-	2.2×10^{10}	7	[47]
	3.7×10^{10}	13	[47]
	3×10^{10}	—	[61]
Ni^{2+}	2.3×10^{10}	—	[64]
NO	3.1×10^{10}	7	[46]
NO_2^-	3.2×10^9	—	[65]
	4.6×10^9	7	[47]
NO_3^-	1.1×10^{10}	7	[46]
	9.3×10^9	—	[44]
N_2O	8.7×10^9	7	[46]
	9.4×10^9	7	[66]
OH	3×10^{10}	11	[46, 52]
O^-	2.2×10^{10}	13	[52]
O_2	1.9×10^{10}	7	[46]
	2.16×10^{10}	—	[57]
SF_6	1.65×10^{10}	—	[66]
SO_3^{2-}	$< 1.3 \times 10^6$	10	[49]
SO_4^{2-}	$< 10^6$	7	[47, 50]
$S_2O_3^{2-}$	6×10^8	—	[44]
Tl^+	2.8×10^{10}	—	[44]
	3.7×10^{10}	8.5	[49]
UO_2^{2+}	7.4×10^{10}	—	[61]

TABLE 3.2

Rate Constants for e_{aq}^- Reactions with Some Organic Solutes

Reactant	Rate constant, $M^{-1} \sec^{-1}$	pH	Reference
Acetaldehyde	3.5×10^9	11	[46]
Acetamide	4×10^7	5.5–6	[67]
Acetate ion	$< 10^6$	10	[46]
Acetic acid	1.8×10^8	5.4	[46]
Acetone	5.9×10^9	7	[46]
	5.2×10^9	14	[42]
Acrylamide	3.3×10^{10}	—	[44]
	2.1×10^{10}	—	[68]
Adenine	3×10^{10}	6	[69]
Adenosine	1×10^{10}	12	[70]
Aniline	$< 2 \times 10^7$	11.94	[70]
Arabinose	$< 10^7$	—	[45]
Benzene	1.2×10^7	11	[71]
Benzoate ion	3.6×10^9	7	[67]
Benzonitrile	1.6×10^{10}	~11	[72]
Bromoacetate ion	6.2×10^9	~10	[73]
Carbon disulfide	3.1×10^{10}	7.7	[70]
Carbon tetrachloride	3×10^{10}	7	[70]
Chloroacetate ion	1.2×10^9	~10	[73]
Chloroacetic acid	6.9×10^9	1.0–1.5	[9]
Chloroform	3×10^{10}	7	[70]
Cyclohexene	$< 10^6$	11	[71]
Cystamine	4×10^{10}	7.3	[74]
Ethanol	$\leq 4 \times 10^2$	—	[58]
Ethyl ether	$< 10^7$	—	[45]
Formaldehyde	$< 10^7$	7	[46]
Formate ion	2.4×10^4	~11	[75]
Formic acid	1.4×10^8	5	[46]
Fumarate ion	7.5×10^9	13	[70]
Guanidine	2.5×10^8	7	[74]
Methane	$< 10^7$	—	[45]
Methanol	$\leq 4 \times 10^2$	—	[58]
Nitrobenzene	2.8×10^{10}	—	[44]
Oxalate ion	4.8×10^7	7.0–7.7	[76]
	3.4×10^9	2.8–4.0	[76]
Oxalic acid	2.5×10^{10}	1.3	[76]
Phenol	1.8×10^7	6.3–6.8	[77]
Purine	1.7×10^{10}	7.2	[70]
Pyridine	3.7×10^9	—	[44]
Ribose	$< 10^7$	—	[45]
Succinate ion	1.2×10^8	6.0	[78]
	3.1×10^7	10.0	[78]
Tetranitromethane	4.6×10^{10}	6	[33]
Thiourea	2.9×10^9	6.41	[70]
Thymine	1.7×10^{10}	6	[70]
	2.7×10^9	12	[70]
Trichloroacetate ion	6.2×10^9	~10	[70]
Urea	3×10^5	7	[78]

II. The Kinetic Salt Effect as Evidence for the Negative Charge on the Hydrated Electron

The sign and magnitude of the charge on a reacting species can be determined from the change in the experimental rate constant with change in the ionic strength of the solution in which the reaction occurs. This is the so-called primary kinetic salt effect [79]. The mathematical expression relating the rate constant to the ionic strength was derived from the Brønsted model of ionic reactions and the Debye–Hückel theory of ionic solutions. It is most often given in the form

$$\log \frac{k}{k_0} = 1.02\, Z_a Z_b \frac{\mu^{1/2}}{1 + \alpha \mu^{1/2}}$$

Here, k is the reaction rate constant at a given ionic strength (μ) of the solution, k_0 is the rate constant at infinite ion dilution, Z_a and Z_b are the charges of the reacting species a and b, and α is a parameter taking into account the proximity of the species a and b; in water at 25°C the value is close to unity for those ions whose ionic radii are about 3–4 Å. The ionic strength of a solution is a characteristic of the ionic concentration in solutions of strong electrolytes which takes into account their valence. It is calculated from the formula

$$\mu = \tfrac{1}{2} \sum C_i Z_i^2$$

where C_i is the ionic concentration in moles per liter and Z_i is the charge on the ion.

From the above expression, it is seen that when $Z_a Z_b$ is positive, zero, or negative, the corresponding reaction rate constant increases, does not change, or decreases, respectively, with increasing ionic strength. Since $Z_a Z_b$ will be positive, zero, or negative when we have reactions of particles of the same charge sign, of a neutral and a charged particle, and of particles of opposite charge sign, respectively, it suffices to know the charge of one of the reacting particles to conclude from the variation of $\log k/k_0$ with μ whether and to what degree the other particle is charged.

A. Effect of Ionic Strength on the Absolute Rate Constant

A group of researchers at Argonne [53] has studied the dependence of the rate constant of reaction of the hydrated electron with $[Fe(CN)_6]^{3-}$ on the increase in the ionic strength of the solution. The reaction was directly followed by measuring the decrease in the absorption of light at 5780 Å, the wavelength which was shown to be suitable for observation of e_{aq}^-. The conditions were adjusted in such a way that the process observed was pseudo-first-order;

hence, the absolute value of the reaction rate constant was calculated. The ionic strength was adjusted by adding Na_2SO_4, $KClO_4$, or $K_4[Fe(CN)_6]$. The pH of the solution was 10.30–10.47. A scavenger of OH radicals, ferricyanide or methyl alcohol, was always present in the solution. The values of $k_{e_{aq}^- + [Fe(CN)_6]^{3-}}$ were found to increase with increasing ionic strength, which means that the short-lived reacting species must be of the same sign as the ferricyanide ion with which it reacts. Hence, the conclusion is drawn that the hydrated electron is negatively charged. In view of the fact that $Z_{[Fe(CN)_6]^{3-}} = -3$, the straight line which best fits the experimental points also confirms that $Z_{e_{aq}^-} = -1$.

B. Effect of Ionic Strength on Relative Rate Constants, the Experiment of Czapski and Schwarz, and Other Measurements

The first complete experiment which showed that in the case of short-lived reducing species we are dealing with a negatively charged species was that of Czapski and Schwarz [10]. Since at that time, the pulsed technique was still not in general use for direct observation of rapid reactions, the authors followed the effect of ionic strength on relative rate constants, i.e., on k_9/k_8, k_{10}/k_8, and k_{11}/k_8, instead of on the absolute value of reaction rate constant k, as in the study described in the preceding section. Reactions (8)–(11), which the authors used for this purpose, were

$$e_{aq}^- + H_2O_2 \longrightarrow OH + OH^- \tag{8}$$

$$e_{aq}^- + O_2 \longrightarrow O_2^- \tag{9}$$

$$e_{aq}^- + H_3O^+ \longrightarrow H + H_2O \tag{10}$$

$$e_{aq}^- + NO_2^- \longrightarrow NO_2^{2-} \tag{11}$$

The corresponding relative rate constants at infinite dilution were taken as k_0 values:

$$(k_9/k_8)_0 = 2.0, \quad (k_{10}/k_8)_0 = 1.95, \quad (k_{11}/k_8)_0 = 0.34$$

The ratio k_9/k_8 was measured in an air-saturated solution of KBr (10^{-4} M) and H_2O_2 (1.22×10^{-4} M). The ionic strength was varied by adding $MgSO_4 \cdot 7H_2O$ (5×10^{-3} M) and $LiClO_4 \cdot 3H_2O$ (4×10^{-2} M). The value

$$\log \frac{k_9/k_8}{(k_9/k_8)_0} = 0$$

was obtained irrespective of the ionic strength of solution, as is to be expected in the case of the reaction of a molecule with a charged particle ($Z_a Z_b = 0$).

The determination of the ratio k_{10}/k_8 was carried out in air-saturated solutions of KBr (10^{-4} M) and $HClO_4$ (10^{-3} M), in the presence of various

II. THE KINETIC SALT EFFECT

quantities of H_2O_2. To change the ionic strength, only lithium perchlorate (10^{-1} M) was used. Measurements of hydrogen peroxide, and the use of an appropriate kinetic expression in which k_{10}/k_8 was the only unknown, showed that the ratio studied decreases with increasing ionic strength. This dependence of the reaction rate constant on ionic strength exists only in cases where the reacting species are of opposite sign ($Z_a Z_b$ is negative).

For the study of the effect of ionic strength on the ratio k_{11}/k_8, use was made of solutions of potassium nitrite (1.02×10^{-3} M) and H_2O_2 (mainly 1.5×10^{-4} M). Oxygen was flushed out of the solutions with nitrogen. To vary the ionic strength, the following salts were used: $LiClO_4 \cdot 3H_2O$, $Li_2SO_4 \cdot H_2O$, $KClO_4$, $NaClO_4$, $MgSO_4 \cdot 7H_2O$, K_2SO_4, and $LaCl_3 \cdot 6H_2O$. The reaction mechanism, the experiment, and the interpretation are somewhat more complex than in the preceding two cases, but the approach is basically the same. The results obtained clearly show that in this case, as is to be expected when a reaction of species of the same charge sign is in question, the rate constant increases with increasing ionic strength.

All these data are summarized in Fig. 3.4. The straight lines are drawn for

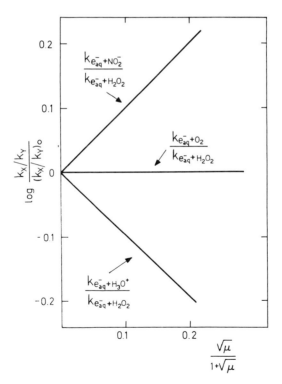

Fig. 3.4. Effect of ionic strength on some relative rate constants of the hydrated electron. (After Czapski and Schwarz [10].)

the slopes $+1.02$ and -1.02, which correspond to unit negative charge on the short-lived reducing species studied, e_{aq}^-. From these data, the authors also concluded that the species in question is thermalized, otherwise the effect of ionic strength could not be followed. It also must move sufficiently slowly to maintain its ionic atmosphere.

By experiments on the effect of ionic strength on relative rate constants, other authors subsequently showed that the hydrated electron has unit negative charge. Collinson et al. [11] have found that the relative rate constant for the reactions

$$e_{aq}^- + Ag^+ \longrightarrow Ag^0 + H_2O \tag{12}$$

$$e_{aq}^- + H_2C:CH \cdot CO \cdot NH_2 \longrightarrow H_3C \cdot \dot{C}H \cdot CO \cdot NH_2 + OH^- \tag{13}$$

decreases with increasing ionic strength of solution at pH 4. Dainton and Watt [80] have found that the relative rate constant for the reactions

$$e_{aq}^- + [Fe(CN)_6]^{3-} \longrightarrow [Fe(CN)_6]^{4-} + H_2O \tag{14}$$

$$e_{aq}^- + N_2O \longrightarrow N_2 + O^- \tag{15}$$

increases with increasing ionic strength of neutral solutions.

III. Evidence for Two Kinds of Reducing Species from Competition-Kinetic Experiments

A. Formation of Molecular Hydrogen and Chloride Ion in the Radiolysis of Degassed Solutions of Chloroacetic Acid

When degassed solutions of monochloroacetic acid are subjected to radiation, H_2, Cl^-, and H_2O_2 appear as the main products of radiolysis. It has been noticed [6] that the yields of the first two products change substantially with change in the concentration of chloroacetic acid (Fig. 3.5).

Since under these conditions the nature of the scavenger does not change, the explanation of the variation in yield lies in the different nature of the reducing species. Depending on scavenger concentration, a reaction giving mainly H_2 and another that produces Cl^- are more or less intense:

$$H + ClCH_2COOH \longrightarrow H_2 + Cl\dot{C}HCOOH \tag{16}$$

$$e_{aq}^- + ClCH_2COOH \longrightarrow Cl^- + \dot{C}H_2COOH \tag{17}$$

In dilute acidic solutions, reaction (10) readily occurs:

$$e_{aq}^- + H_3O^+ \longrightarrow H + H_2O \qquad [10]$$

Hence, reducing species have the form of H atoms, and reaction (16) predominates. As the concentration of chloroacetic acid increases, competition arises between reactions (17) and (10); reaction (17) is the more effective, the

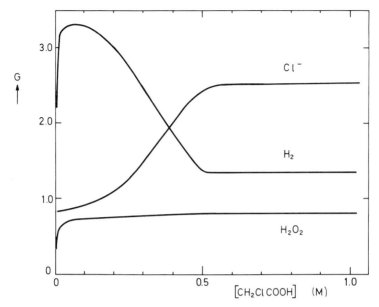

Fig. 3.5. Effect of chloroacetic acid concentration on the yields of products formed in irradiated aqueous solutions (pH = 1) of $CH_2ClCOOH$. (After Hayon and Weiss [6].)

higher the concentration of scavenger. At the same time, the effectiveness of reaction (16) decreases and thereby the yields of molecular hydrogen are lowered.

The work of Hayon and Allen [9] showed in a yet more convincing way that in fact competition between reactions (17) and (10) is in question. At a constant scavenger concentration the increase of pH from acidic to neutral, that is, inhibition of reaction (10), gives rise to a change in the yield, just as an increase in scavenger concentration does at a constant pH (Fig. 3.6).

It is in these experiments that the ratio k_{10}/k_{17} was established to be 3.4. From the ensemble of results, the authors concluded that in irradiated water there are two kinds of reducing species and that one is the precursor of the other and may be converted into this by reaction with the hydrogen ion. Likewise, it is noteworthy that the yields of molecular hydrogen are also measurable at those pH's at which reaction (10), and hence also the production of molecular hydrogen by reaction (16), are practically negligible. Since here the $G(H_2)$ measured is higher than the yield of primary hydrogen (G_{H_2}), the contribution of reaction (16) was explained by the presence of the H atoms originating from the primary act of water radiolysis rather than by the occurrence of reaction (10).

Chloroacetic acid is a suitable scavenger for following the competition between H_3O^+ and the hydrated electron, since it gives different products

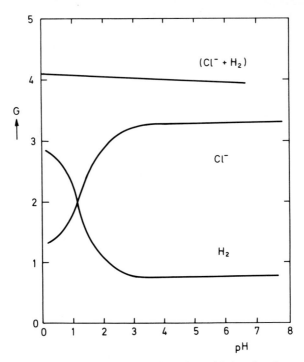

Fig. 3.6. Effect of pH on the yields of products formed in irradiated aqueous solutions of $CH_2ClCOOH$ (0.1 M). (After Hayon and Allen [9].)

and reacts sufficiently rapidly with the solvated electron and the hydrogen atom. The value for k_{17} is about 7×10^9 M^{-1} sec^{-1} and for k_{16} it is a thousand times lower [40]. True, not all the stable products of these reactions are known, and the material balance is far from being complete (under certain conditions the following products are encountered: formaldehyde, glyoxylic acid, and, at higher concentrations, CO). It should also be recalled that the reaction with H atoms can produce chloride ions and that the rate constant for this reaction is only ten times lower than for molecular hydrogen formation by reaction (16).

B. Formation of Hydrogen Peroxide in Irradiated Aqueous Solutions Containing Oxygen and Hydrogen

The study of the radiolysis of aqueous solutions of oxygen, hydrogen, and hydrogen peroxide [81] has contributed significantly to our understanding of what happens to water as radiation passes through it. Owing to the relatively simple reaction schemes involved and to the possibility of precise determination of the radiolysis products, competition-kinetic studies have provided, among other insights, very convincing evidence for the existence of two kinds

IV. ATOMIC HYDROGEN

of reducing species. Barr and Allen [7] have studied the formation of hydrogen peroxide in hydrogen-saturated water in which oxygen was present at different initial concentrations. The authors have noticed that the H atom produced in

$$OH + H_2 \longrightarrow H + H_2O \qquad (18)$$

reacts considerably faster with O_2 than with H_2O_2. In solutions in which there is no molecular hydrogen and no source of H atoms given by the above reaction, but only those reducing species are present which arise from water radiolysis (at neutral pH), the observed rate constants for reactions of short-lived reducing species with O_2 and H_2O_2 are practically identical. From this the conclusion was drawn that two kinds of short-lived reducing species are in question. Today, we can easily calculate that for the reactions observed in the above case,

$$e_{aq}^- + O_2 \longrightarrow O_2^- \qquad [9]$$
$$H + O_2 \longrightarrow HO_2 \qquad (19)$$
$$e_{aq}^- + H_2O_2 \longrightarrow OH + OH^- \qquad [8]$$
$$H + H_2O_2 \longrightarrow OH + H_2O \qquad (20)$$

the following ratios hold:

$$\frac{k_{e_{aq}^- + O_2}}{k_{e_{aq}^- + H_2O_2}} \sim 1.6$$

whereas

$$\frac{k_{H + O_2}}{k_{H + H_2O_2}} \sim 200$$

IV. ATOMIC HYDROGEN

A. Its Role in the Radiolysis of Aqueous Solutions Is Considerably Smaller Than Was Initially Supposed

As we saw in Chapter Two, for a long time the water radiolysis model considered the hydrogen atom as the only reducing species. The experiments described in Section III of this chapter revealed that two kinds of reducing species are present in irradiated water. A vast amount of information that followed these experiments showed that the role of H atoms in the radiolysis of aqueous solutions is considerably less than that of hydrated electrons. At present, it is evident that this situation results from the facts that, at the beginning of the chemical stage of radiolysis, the amount of H atoms is substantially less than that of e_{aq}^- (very likely by a factor of 5), and that their reactivity is in general considerably weaker. This can be seen from Table 3.3, which presents a comparison of rate constants for some of the more important reactions of the two species.

TABLE 3.3

Comparison of Rate Constants for e_{aq}^- and H Reactions with Some Solutes and Free Radicals

Reactant	Rate constant, M^{-1} sec^{-1}		pH	Reference
	e_{aq}^-	H		
OH	3×10^{10}	—	11	[52]
	—	3.2×10^{10}	0.4–3	[82]
H_3O^+	2.36×10^{10}	—	4–5	[53]
	—	2.6×10^3	—	[83]
H_2O_2	1.23×10^{10}	—	7	[53]
	—	9×10^7	2	[84]
O_2	1.9×10^{10}	—	7	[46]
	—	1.9×10^{10}	2	[84]
CO_2	7.7×10^9	—	7	[46]
	—	$< 10^6$	5	[85]
N_2O	9.4×10^9	—	7	[66]
	—	1.25×10^4	—	[83]
NO_3^-	1.1×10^{10}	—	7	[46]
	—	9×10^6	7	[86]
Cu^{2+}	3.4×10^{10}	—	1.5–4.5	[51]
	—	4.2×10^7	1.5	[51]
Fe^{2+}	3.5×10^8	—	—	[50]
	—	1.6×10^7	0.4	[87]
Acetone	5.9×10^9	—	7	[46]
	—	3.5×10^5	1	[88]
Ethanol	$\leq 4 \times 10^2$	—	—	[58]
	—	1.32×10^7	2	[89]
Formate ion	2.4×10^4	—	11	[75]
	—	2.6×10^8	1	[90]
Formic acid	1.4×10^8	—	5	[46]
	—	1.1×10^6	1	[90]
Methanol	$\leq 4 \times 10^2$	—	—	[58]
	—	1.56×10^6	2	[89]
Tetranitromethane	4.6×10^{10}	—	6	[33]
	—	2.6×10^9	—	[91]

In addition to relatively low yields and reactivity, a difficulty encountered in studying the nature of H atoms, as in the case of other radical products of radiolysis, is the fact that they do not appear alone. This is why the experiments with H atoms produced by microwave discharge in gaseous molecular hydrogen and introduced into various solutions are of special interest [86,92–95]. Experimental details on this method are given in Chapter Seven, together with some other techniques employed in radiation chemistry. This technique made it possible to have only H atoms as the reactive primary species in solution. Within the limits of experimental error, the kinetic data were in

IV. ATOMIC HYDROGEN

agreement with results obtained, for example, by studying the γ-radiolysis of ferrous sulfate. In view of the fact that the concentration of reactive species is considerably higher around the gas bubble at the inlet to the solution, remarks as to whether homogeneous kinetics can be applied to such cases have been discussed in detail [93]. A comparison of the rate constant values obtained in this way with those from radiation-chemical experiments shows considerable discrepancies; rate constants derived from the gas-discharge technique assuming homogeneous kinetics may be an order of magnitude smaller than the true ones. However, the diffusion model and heterogeneous kinetics give a satisfactory answer [86].

It should be noted that under certain circumstances concentrations of H atoms may be considerable and even higher than those of hydrated electrons. This is the case in acidic solutions in which

$$k_{e_{aq}^- + H_3O^+}[H_3O^+] \gg k_{e_{aq}^- + S}[S]$$

where S is the solute (or solutes) reacting with the hydrated electron. Certain acids in reaction with e_{aq}^- also give H atoms [96]. The values of the rate constants for the reactions

$$e_{aq}^- + HA \longrightarrow H + A^- \tag{21}$$

may be rather high (about $10^8\ M^{-1}\ \text{sec}^{-1}$), as in the case of acetic, phosphoric, and formic acids [97]. One of the possibilities of increasing the concentration of H atoms in the solution studied, which has already been mentioned, is conversion of the hydroxyl radical by reaction with H_2 [Eq. (18)].

B. Interconversion of H and e_{aq}^- on the H_2^+ Species

The inevitable presence of H_3O^+ ions in water makes it necessary to take into account the reaction

$$e_{aq}^- + H_3O^+ \longrightarrow H + H_2O \tag{10}$$

Although this represents the replacement of one short-lived reducing species by another, it is nevertheless not negligible for chemical processes in the irradiated medium. As we have seen in the preceding section, the hydrogen atom often reacts at different rates and also may give chemically different products from the hydrated electron. Hence, it is necessary to have an accurate value for the rate constant for the above reaction.

Since the concentration of e_{aq}^- may be directly measured spectrophotometrically, the absolute value of the rate constant for reaction (10) can be simply measured. We shall dwell on this at some length, because it represents a suitable example of a pseudo-first-order reaction of a type frequently and very successfully used for measurements of absolute rate constants.

Let us first consider the general case of the pseudo-first-order process. Suppose radical R reacts with solute S to give product P

$$R + S \longrightarrow P$$

To measure the rate constant, it is necessary to follow either the production of P or the disappearance of R. Let us assume that the latter has a suitable absorption spectrum in a region where P does not absorb, so that the reaction rate is conveniently followed through R,

$$-\frac{d[R]}{dt} = k[R][S]$$

If we adjust the experimental conditions in such a way that the concentration of reacting substance S is sufficiently high so that it remains practically constant during the course of observations, i.e.,

$$[S] \gg [R]$$

then we may write $k' = k[S]$, and substituting the new constant into the differential equation above, we obtain the expression for the pseudo-first-order process

$$-\frac{d[R]}{dt} = k'[R]$$

which is more conveniently written as

$$-\frac{d[R]}{[R]} = k' \, dt$$

where k' is the pseudo-first-order rate constant (in reciprocal seconds) from which we may, knowing the concentration of the solute, calculate the unknown rate constant k (in liters per mole per second) for the second-order process,

$$k = \frac{k'}{[S]}$$

Integrating the differential equation in the range from $t = 0$ to $t = t$ gives

$$\ln \frac{[R]_0}{[R]_t} = k't$$

or, which amounts to the same thing but is more convenient for graphical representation,

$$\log \frac{OD_0}{OD_t} = 0.434 \, k't$$

IV. ATOMIC HYDROGEN

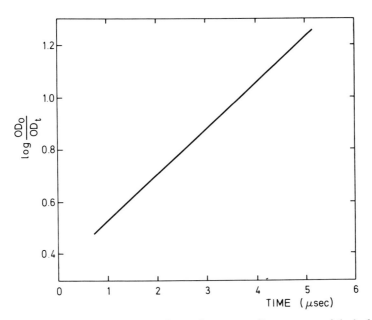

Fig. 3.7. An example for the pseudo-first-order process: disappearance of the hydrated electron by the reaction with the hydrogen ion. (After Dorfman and Taub [55].)

where OD is the optical density of the radical R measured at a convenient wavelength.

Figure 3.7 presents such a case. Experimental details are taken from Dorfman and Taub [55]. The authors have observed the decay kinetics of the hydrated electron at 5770 Å following a 0.4-μsec pulse. The solutions contained 0.5 M ethanol so that the hydroxyl radical was effectively removed and did not disturb the observation of reaction (10). The range of hydrogen ion concentration and the pulse current were selected so that $[H_3O^+]/[e_{aq}^-] \gg 10$. Under these conditions the process observed is pseudo-first-order with respect to the hydrated electron. The measurement could be readily carried out because the high molar extinction coefficient of the e_{aq}^- allows observations to be performed at very low concentrations of this species. Figure 3.7 shows the test of the first-order rate law for the disappearance of the hydrated electron in these solutions where $[H_3O^+] = 1.9 \times 10^{-5}$ M. The logarithm of the measured optical densities is plotted as ordinate and the time after pulse as abscissa. From the slope of the straight line, one calculates

$$\tan \alpha = 0.434 \, k' = \left[\left(\log \frac{OD_0}{OD_t} \right)_{t_2} - \left(\log \frac{OD_0}{OD_t} \right)_{t_1} \right] / (t_2 - t_1)$$

$$= (1.210 - 0.534)/(5 \times 10^{-6} - 1 \times 10^{-6}) = 0.169 \times 10^6$$

that is,

$$k' = 0.4 \times 10^6 \quad \sec^{-1}$$

Whence we obtain the absolute value:

$$k = \frac{k'}{[\text{H}^+]} \sim \frac{0.4 \times 10^6 \quad \sec^{-1}}{1.90 \times 10^{-5} \quad M} \sim 2.1 \times 10^{10} \quad M^{-1} \sec^{-1}$$

The value derived from Fig. 3.7 is only one of a number of determinations, pH 4.1–4.7 at 23°C, for which the authors give the final result:

$$k_{e_{aq}^- + \text{H}_3\text{O}^+} = (2.26 \pm 0.21) \times 10^{10} \quad M^{-1} \sec^{-1}$$

The values obtained by other authors are given in Table 3.1.

As compared to the fast conversion of e_{aq}^- into H, the conversion of hydrogen atoms into hydrated electrons,

$$\text{H} + \text{OH}_{aq}^- \longrightarrow e_{aq}^- + \text{H}_2\text{O} \tag{22}$$

is quite slow, the value for k_{22} being a thousand times lower than k_{10}. However, this reaction is very important, especially in alkaline solutions.

The existence of the reaction of the H atoms with hydroxyl ions was first indisputably shown by experiments of Jortner and Rabani [98], in which the atomic hydrogen produced by microwave discharge was introduced into solutions of monochloroacetic acid at various pH's. As we have seen in Section IIIA, only the reaction with the hydrated electron can lead to the formation of the chloride ion:

$$e_{aq}^- + \text{CH}_2\text{ClCOOH} \longrightarrow \text{Cl}^- + \dot{\text{C}}\text{H}_2\text{COOH} \qquad [17]$$

As a matter of fact, up to pH 10 the amount of the chloride ions is negligible (a few percent). It increases as pH increases up to 12.5, when practically each H atom introduced generates a Cl^- ion. The explanation of this lies in the reaction of the H atoms with OH^- ions to give hydrated electrons, which then react efficiently with monochloroacetic acid to give Cl^- [Eq. (17)].

Elegant evidence for the conversion was obtained by Matheson and Rabani [52] by direct measurement of light absorption at the wavelength characteristic of the hydrated electron. Experimental conditions were adjusted in such a way that, if the conversion does exist, it must be accompanied by a corresponding increase or decrease in the absorption. The concentration of molecular hydrogen was about 0.1 M, and this was achieved in a special cell using a pressure of hydrogen gas up to 100 atm. Hence, all OH radicals were first converted into H atoms by the reaction

$$\text{OH} + \text{H}_2 \longrightarrow \text{H} + \text{H}_2\text{O} \qquad [18]$$

and then the H atoms produced were transformed into e_{aq}^- according to

IV. ATOMIC HYDROGEN

reaction (22). The authors found that as the pH increased from 7 to 11, no change occurred in the value of the measured initial absorption. But in the pH range 12–13, a substantial increase occurred. In alkaline media, where the conversion of H atoms is efficient and where these arise not only from the primary act of radiolysis but also from the conversion by hydroxyl radicals, the ratio of the measured absorption to that observed in neutral solutions must be the same as the ratio $(G_H + G_{OH} + G_{e_{aq}^-})/G_{e_{aq}^-}$. The value obtained experimentally is 2.2, in good agreement with that calculated on the basis of primary yield data (see Chapter Five). Some other observations made in the course of these experiments provided further evidence for the conversion. Thus, for example, the absorption observed after the pulse at pH 11.6 due to the hydrated electron first increases with time, then decreases, and finally falls to zero. This can only be explained by assuming that the initial absorption is due to the hydrated electrons, corresponding quantitatively to $G_{e_{aq}^-}$, and that the increase results from the conversion of OH radicals into H atoms and of these last into hydrated electrons [Eqs. (18) and (22)].

At low pH, where the conversion is negligible (for example, at a pH of about 10), the absorption begins to decrease immediately after the pulse. The same effect is also observed at pH 13.3, but for a different reason: Since the concentration of OH^- is sufficiently high, conversion of OH and H into e_{aq}^- is already completed in the course of the pulse duration. Hence, the reaction $H + OH^-$ may be expected to be best observed in moderately alkaline solutions at moderate dose rates, and with the other conditions adjusted in such a way that the second-order reactions by which e_{aq}^- disappears are of minor importance. Such conditions were found at pH 11.6 and $k_{22} = (1.8 \pm 0.6) \times 10^7 \, M^{-1} \, \text{sec}^{-1}$. This value is somewhat lower than those reported in other publications, e.g., $2.2 \times 10^7 \, M^{-1} \, \text{sec}^{-1}$ [97]. While discussing the conversion of H into e_{aq}^-, it is interesting to consider the observation of Anbar and Neta [99]. Starting from the fact that the fluoride ion has similar valence-electron structure with OH^-, the authors assume that the following reaction of conversion, analogous to that of the H atom with OH^-, is also possible:

$$H + F^- \longrightarrow HF^- \longrightarrow HF + e_{aq}^- \qquad (23)$$

This conversion has been demonstrated by the increase in the yield of electrons in the radiolysis of neutral solutions, as well as in sonolytic experiments, where e_{aq}^- are initially absent. The rate constant for this reaction is fairly low, $k_{23} = 1.5 \times 10^4 \, M^{-1} \, \text{sec}^{-1}$.

In Section I.C, we considered the reaction of the hydrated electron with water to give the H atom [Eq. (6)]. Here, we have considered the reverse reaction, the conversion of the hydrogen atom into e_{aq}^- [Eq. (22)]. The known values for k_6 and k_{22} allow us to calculate the equilibrium constant for the reaction

$$\text{H} + \text{OH}^- \underset{k_6}{\overset{k_{22}}{\rightleftarrows}} \text{e}_\text{aq}^- + \text{H}_2\text{O}$$

that is, pK = 9.6 for the H atom. It can be seen that it is a very weak acid. In strongly acid media one has to take into account the reaction

$$\text{H} + \text{H}_3\text{O}^+ \longrightarrow \text{H}_2^+ + \text{H}_2\text{O} \tag{24}$$

although this is very slow; reported values for k_{24} vary from 0.5×10^2 to $2.6 \times 10^3 \, M^{-1} \, \text{sec}^{-1}$ [83,100]. During a long period, the H_2^+ species was thought to play an important role in the oxidation of ferrous ions in acidic solutions ($\text{H}_2^+ + \text{Fe}^{2+} \rightarrow \text{Fe}^{3+} + \text{H}_2$) [87,92,101]. However, it has been shown [102] that the oxidation yields of Fe^{2+} do not depend on the H_3O^+ concentration when it is higher than 1 M, and that the H_2^+ species does not play any considerable role even in this reaction. It seems that oxidation by hydrogen atoms in acid aqueous solutions is a termolecular reaction ($\text{H} + \text{H}^+ + \text{Fe}^{2+} \rightarrow \text{Fe}^{3+} + \text{H}_2$), as shown for the oxidation of the iodide ion [103a]. Nevertheless, Hentz et al. [103b] explain their results with two steps: $\text{Fe}^{2+} + \text{H} \rightarrow \text{FeH}^{2+}$ and $\text{FeH}^{2+} + \text{H}^+ \rightarrow \text{Fe}^{3+} + \text{H}_2$.

C. Reactivity

We have seen that the hydrated electron reacts with water very slowly ($k_6 = 16 \, M^{-1} \, \text{sec}^{-1}$). No corresponding reaction, however, was found to involve the hydrogen atoms, as pointed out by the absence of the exchange reaction in reactions of D atoms [104].

The reducing properties of the H atom are demonstrated in a number of reactions with inorganic ions as, for example, in the case of divalent copper,

$$\text{Cu}^{2+} + \text{H} \longrightarrow \text{Cu}^+ + \text{H}^+ \tag{25}$$

The case of abstraction of hydrogen from organic molecules in the reaction producing molecular hydrogen,

$$\text{H} + \text{HR} \longrightarrow \text{H}_2 + \text{R} \tag{26}$$

is also frequent. The rate constants for reactions of the H atom have been studied considerably less than those of the hydrated electron. In certain cases, atomic hydrogen produced by microwave discharge was used for direct measurements [86,95,105]. As in the case of other primary radicals, competition kinetics was of great use in determining the rate constants [90,106–110]. Also, experiments with accelerated electrons, pulsed-beam, or steady continuous irradiation have furnished valuable information [82,84,111–114]. Some rate constants are presented in Tables 3.4 and 3.5. One of the reasons for the scarcity of information on these rate constants is certainly due to the fact that the hydrogen atom, unlike e_aq^-, has no characteristic spectrum suitable for routine measurements. Pulse-radiolytic experiments [116,126] with

TABLE 3.4

Rate Constants for H Atom Reactions with Some Inorganic Solutes and Free Radicals

Reactant	Rate constant, $M^{-1} \text{ sec}^{-1}$	pH	Reference
Ag^+	1.15×10^{10}	7	[115]
Cd^{2+}	$< 10^5$	7	[108]
CO_2	$< 10^6$	5	[85]
CrO_4^{2-}	3.4×10^9	7	[108]
$Cr_2O_7^{2-}$	5.6×10^9	7	[108]
Cu^{2+}	4.2×10^7	1.5	[51]
	6.4×10^7	1	[88]
e_{aq}^-	2.5×10^{10}	10.5	[52]
	3×10^{10}	10.9	[46]
Fe^{2+}	1.6×10^7	0.4	[87]
	1.3×10^7	2.1	[87]
	2×10^7	0.2–0.8 N H_2SO_4	[111]
Fe^{3+}	1×10^6	0.4	[87]
	9.4×10^7	2.1	[87]
$Fe(CN)_6^{3-}$	8.7×10^9	1.71	[90]
	6.5×10^9	2–3	[114]
H	1.5×10^{10}	0.2–0.8 N H_2SO_4	[111]
	1.0×10^{10}	2	[84]
	7.75×10^9	3	[116]
	1.3×10^{10}	0.4–3	[82]
	1.25×10^{10}	2–3	[114]
HCO_3^-	$\sim 3 \times 10^4$	7	[117]
H_2O_2	5×10^7	Acid	[118]
	1.6×10^8	0.4–3	[82]
	4×10^7	—	[112]
	9×10^7	2	[84]
H_3O^+	2.6×10^3	—	[83]
	$5 \times 10^1 – 5 \times 10^2$	—	[100]
HNO_2	1×10^9	1	[119]
HPO_3^{2-}	1.8×10^9	13	[120]
MnO_4^-	2.8×10^{10}	3	[61]
N_2O	1.25×10^4	—	[83]
	$\sim 10^3 – 10^4$	2	[121]
	$\sim 10^5$	—	[122]
NO_2^-	2.4×10^9	8	[119]
NO_3^-	7×10^6	11–13	[117]
	9.3×10^6	7	[86]
	1.2×10^7	1.1	[86]
O_2	1.2×10^{10}	Acid	[112]
	2.6×10^{10}	0.4–3	[82]
	1.9×10^{10}	2	[84]
OH	3.2×10^{10}	0.4–3	[82]
OH^-	1.8×10^7	11.5	[52]
	2.2×10^7	11–13	[97]
Zn^{2+}	$< 10^5$	7	[108]

TABLE 3.5

RATE CONSTANTS FOR H ATOM REACTIONS WITH SOME ORGANIC SOLUTES

Reactant	Rate constant, $M^{-1}\ \text{sec}^{-1}$	pH	Reference
Acetate ion	1×10^6	7	[123]
Acetic acid	1×10^5	1	[88]
Acetone	3.5×10^5	1	[88]
Aniline	1.8×10^9	7	[124]
Benzene	5.3×10^8	2	[71]
Benzoate ion	8.7×10^8	7	[124]
Benzoic acid	1×10^9	1	[125]
Benzonitrile	6.8×10^8	1	[125]
	4.5×10^8	7	[124]
Cyclohexene	3×10^9	2	[71]
Ethanol	1.5×10^7	1	[88]
	1.7×10^7	1–3	[90]
	1.32×10^7	2	[89]
Formaldehyde	5×10^6	1	[88]
Formate ion	2.6×10^8	1–3	[90]
	2.5×10^8	11–13	[117]
Formic Acid	1.1×10^6	1	[90]
Glycerol	1.45×10^7	2	[89]
Glycine	1.71×11^7	2	[89]
Glycol	7.6×10^6	2	[89]
Methanol	1.56×10^6	2	[89]
	1.6×10^6	1	[88]
	1.6×10^6	1	[125]
Nitrobenzene	1×10^9	1	[125]
2-Propanol	3.9×10^7	2	[89]
1-Propanol	1.44×10^7	2	[89]
Sucrose	3.8×10^7	1	[88]
Tetranitromethane	2.6×10^9		[91]

deaerated solutions of $10^{-3}\ M$ HClO$_4$ + 0.027 M H$_2$ have shown the existence of a species absorbing in the region between 2000 and 2400 Å. The authors suggest that this spectrum belongs to the hydrogen atom and give $\varepsilon_{2000} = 900\ M^{-1}\ \text{cm}^{-1}$ and $\varepsilon_{2400} = 0\ M^{-1}\ \text{cm}^{-1}$. The assumption was made on kinetic grounds; transient decays were studied under various conditions and the experimental data were found to agree with the well-established values for the hydrogen atom. However, some doubt still remains because of the strong shift of the observed spectrum towards the red; in the gas phase hydrogen atoms do not absorb light in the ground state at wavelengths longer than 1215 Å.

IV. ATOMIC HYDROGEN

D. Origin of Primary Atomic Hydrogen

It is generally accepted that $G_H \approx 0.6$, independently of pH [8,117,127]. We have seen in Chapter Two that in interaction with radiation, a certain number of water molecules undergo excitation that may lead to their dissociation:

$$H_2O^* \longrightarrow H + OH \qquad (27)$$

Consequently, primary atomic hydrogen arises from this reaction. Another possible origin of the H atoms may be the intraspur reaction between the protonated water molecule and the hydrated electron [128,129]:

$$e^-_{aq} + H_3O^+ \longrightarrow H + H_2O \qquad [10]$$

For a good agreement to be obtained between calculated and experimental values, diffusion-kinetic calculations point to the importance of taking into account reaction (10). The question arises, of course, as to whether the precision of the calculations is sufficient to take into account the effect, which certainly is not very large. The experiment which should give the answer is, in principle, quite simple: Increasing the concentration of an efficient scavenger for e^-_{aq} or H_3O^+ should reduce reaction (10) and hence decrease G_H. As will be seen from the contradictory answers, the situation seems to be more complicated; the use of high scavenger concentrations makes the intraspur reactions more complex than is generally admitted. One should also know for certain that under such conditions the scavenger has no effect on H_2O^* deactivation, either because no energy transfer occurs when a particular solute is in question, or because the half-life of H_2O^* is very short.

In some experiments [130], ethanol (up to 10 M), isopropanol, or formate (up to 1 M) were used as H atom scavengers. Hydrated electrons were scavenged by nitrate, the concentration of which was varied up to 2 M. Measurements were carried out in neutral solutions at pH's ranging from 5.7 to 7.1, but subsidiary experiments showed that, in certain cases, identical results are also obtained at pH 9–10. The experiments showed that the higher the nitrate concentration, the larger the decrease in the value of G_H as derived from competition plots. The data are presented in Fig. 3.8. We see that when the nitrate concentrations lie between 10^{-6} and 10^{-3} M, G_H is approximately 0.7 and thereafter decreases with increasing nitrate concentration. In a 2 M solution, G_H is only about 0.3. From these results the authors concluded that a considerable part of G_H is due to the intraspur reaction of hydrated electrons with H_3O^+ [Eq. (10)]. They think that the existence of another source of hydrogen atoms, such as the process given by Eq. (27) (that is, dissociation of H_2O^*) is not thereby ruled out, but conclude that its role must be much less important than previously supposed.

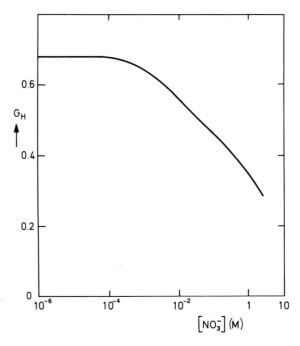

Fig. 3.8. Effect of nitrate concentration on the primary hydrogen atom yield. The data were derived from measurements on irradiated neutral aqueous solutions of nitrate containing ethanol, isopropanol, or formate. (After Chouraqui and Sutton [*130*].)

Appleby [*131*] has also shown how the increase in concentration of an efficient e_{aq}^- scavenger leads to a drop in G_H, concluding that reaction (10) is the main, although not exclusive, source of primary atomic hydrogen yield. Acetone served as the scavenger for e_{aq}^- and isopropanol was used as the hydrogen atom scavenger. It was shown how the increase in acetone concentration leads to a decrease in G_{H_2} and G_H. Both yields, when represented as a function of the cube root of concentration, behave similarly. One can practically draw two parallel lines through the experimental data; the intercepts on the ordinates give $G_{H_2}^0 = 0.45$ and $G_H^0 = 0.71$. The variation in yield is expressed as a function of the cube root of solute concentration, $G/G^0 = 1 - (kS)^{1/3}$, where G^0 is the yield at infinite dilution and S is the solute concentration. If several solutes are present, as was the case in some experiments, kS is the sum of solute concentrations multiplied by the respective rate constants. The author concluded that the resemblance in the behavior of G_H and G_{H_2} data implies strongly that a similar mechanism is operating in both cases, i.e., that the solute reacts in the spur with entities which would otherwise combine during diffusion. He also points out that there is a tendency for G_H

to lie somewhat above G_{H_2}, which may suggest that only a portion of the H atom yield has precursors scavengeable by the solutes used.

Instead of studying the existence of reaction (10) in the spur by scavenging hydrated electrons, it is possible to approach the problem by capturing the hydrogen ion

$$OH_{aq}^- + H_3O_{aq}^+ \longrightarrow 2H_2O \qquad [5]$$

In principle, this process should also, under suitable conditions, lead to a decrease in G_H and hence to a decrease in molecular hydrogen production by the reaction $H + HR \rightarrow H_2 + R$. Such experiments have been carried out, and the molecular hydrogen yields have been measured at various pH in solutions containing low concentrations of solutes RH and of nitrate ion (RH = methanol, ethanol, formate, or phosphite) [132]. The results obtained indicate that all the data up to pH 13.5 can be explained in terms of the competition between RH and OH^- for a constant hydrogen atom yield, $G_H = 0.50 \pm 0.05$, which means that reaction (10) is not the source of primary atomic hydrogen.

The isotopic effect was studied in reactions that are assumed to lead to the formation of molecular hydrogen: $e_{aq}^- + e_{aq}^-$, $e_{aq}^- + H$, $H + H$ [133]. The measurements were concerned with the abundance of H in the products (H_2, H, OH, H_2O_2) of the radiolysis of mixtures of H_2O and D_2O. The kinetic isotope effect was defined as the ratio of the fraction of H atoms in the product to that in the water, and it constitutes a measure of the difference in the rates of processes leading to bond rupture, that is, of the reaction assumed to produce the radiolytic product studied ($H_2O \leadsto H + OH$, $H_3O \rightarrow H + H_2O$). The conclusion was that atomic hydrogen most likely originates from excited water molecules [Eq. (27)] and not from reaction (10), as assumed in the diffusion-kinetic model. However, it has been shown by a detailed calculation that under the above experimental conditions the isotopic effect must be very small (4% and less) and that its absence cannot be taken as evidence against reaction (10) in the spur being the source of H atoms [134].

Some experiments tried to answer the question as to whether the reducing species that was assumed to be the H atom is in fact the H atom, or a long-lived excited water molecule [135]. The assumption was that reactions of long-lived H_2O^* might become important in solutions containing high concentrations of scavengers. If this is so, then reaction rate constants measured in concentrated solutions would be different from those measured in dilute solutions. The results of the measurements show that, within the limits of experimental error, there are no differences. Moreover, the values obtained are of the same order of magnitude as those found using atomic hydrogen produced in a gas discharge. Hence, the conclusion follows that we are dealing with a H atom produced 01^{-8} sec or less after the passage of the radiation, and not with an excited water molecule whose lifetime would be equal to or longer than 100 μsec.

V. Primary Molecular Hydrogen

A. Decrease in G_{H_2} with Increasing Concentration of e_{aq}^- Scavenger

The conclusion that $G_{H_2} \simeq 0.45$, irrespective of pH (see Chapter Five), seems to be more or less beyond doubt. Its decrease with increasing concentration of e_{aq}^- scavengers gives information about the way in which this primary molecular hydrogen is produced. From the earlier sections of this chapter it is evident that e_{aq}^- represents the major part of the short-lived reducing species, and that the origin of the primary molecular hydrogen may be sought, in the first instance, in the recombination,

$$e_{aq}^- + e_{aq}^- \xrightarrow{2H_2O} H_2 + 2OH^- \qquad (28)$$

and, to a considerably less extent, in the reaction,

$$e_{aq}^- + H \xrightarrow{H_2O} H_2 + OH^- \qquad (29)$$

The contribution of the recombination of H atoms,

$$H + H \longrightarrow H_2 \qquad (30)$$

must be still smaller. If these reactions are indeed the only source of G_{H_2}, then a sufficient increase in concentration of an efficient scavenger (S) for hydrated electrons must also lead to an effective decrease in G_{H_2}. This was actually found to be the case with various solutes: hydrogen peroxide [136,137], potassium nitrite [138], potassium or sodium nitrate [139–141], copper sulfate [138], copper nitrate [139], ceric sulfate [142], and potassium dichromate [139]. The effect of nitrate concentration on the measured yields G_{H_2}, as well as on the simultaneously derived yields G_H, is illustrative [130]: as the nitrate concentration increases from 10^{-6} to $2\,M$, in solutions containing alcohol as a scavenger for OH and H, a considerable decrease in both G_{H_2} and G_H is observed. For $[NO_3^-] = 2\,M$, G_{H_2} is reduced to 12% of its value in dilute solution (to about 0.054), whereas G_H is diminished by only about 50%. This points to the importance of the recombination of hydrated electrons in the production of molecular hydrogen, as well as to the importance of the intraspur reaction $e_{aq}^- + H^+$ as a source of H atoms. From the fact that practically all molecular hydrogen disappeared, whereas atomic hydrogen did not, the authors concluded that the dissociation of excited water molecules may be a source of primary H atoms but not of H_2 molecules.

These results have shown that G_{H_2} is dependent on the concentration of the e_{aq}^- scavenger present in the solution studied. However, they have not supplied enough information for a quantitative correlation to be established between

V. PRIMARY MOLECULAR HYDROGEN

the measured G_{H_2} and the scavenger reactivity, $k_{e_{aq}^- + S}[S]$. The fact that some scavengers characterized by similar rate constants with the hydrated electron reduced G_{H_2} with quite different efficiencies was especially puzzling. The work of Peled and Czapski [48] shows that these difficulties are mainly due to ionic strength effects on the rate constants of the hydrated electron with the scavengers used. In their work, G_{H_2} was measured in the γ-radiolysis of air-free solutions (0.01, 0.1, and 1 M) of various cations, anions, and neutral molecules known as efficient e_{aq}^- scavengers. Also, the rate constants of most of these substances at different ionic strengths were directly determined by following e_{aq}^- decay in pulsed electron beam experiments. The results obtained show that, if the values of $k_{e_{aq}^- + S}$ are taken for $\mu = 0$, for each group (cations, anions, neutral molecules) there is a satisfactory correlation between the measured G_{H_2} and the reactivity. However, when the three groups are compared, a difference is seen: For the same reactivity, anions are more efficient in decreasing molecular hydrogen yield than the other two groups (Fig. 3.9). For example, the rate constants of e_{aq}^- with NO_3^-, H_2O_2, and Co^{2+} are about 10^{10} M^{-1} sec^{-1}; the G_{H_2} values recorded in 1-M solutions were 0.094, 0.19, and 0.27, respectively. A unified curve is obtained when the rate constants are chosen so as to correspond not to $\mu = 0$ but rather to the ionic strength of the solution in which the particular hydrogen yield was measured. This curve corresponds mainly to the curve given above for neutral

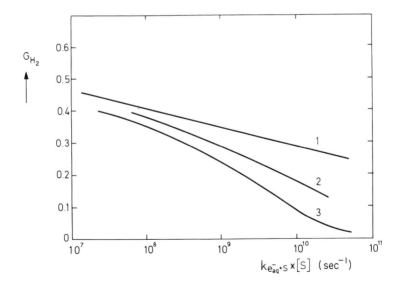

Fig. 3.9. Effect of reactivity ($k_{e_{aq}^- + S} \times [S]$ sec^{-1}) on the primary molecular hydrogen yield: cationic scavengers (curve 1), uncharged scavengers (curve 2), and anionic scavengers (curve 3). (After Peled and Czapski [48].)

molecules, that is, for the scavenging when the rate constants are not affected by ionic strength.

These data provide strong evidence that the precursor of G_{H_2} is a negatively charged species and that the reactions of e_{aq}^- [Eqs. (28) and (29)] are of importance in accounting for the origin of the primary molecular hydrogen. It was also shown that in concentrated solutions, G_{H_2} can be decreased to about 5% of its value in dilute solutions. This leads to the conclusion that almost all precursors of G_{H_2} may be scavenged and that there is no direct formation of primary molecular hydrogen. The finding that not all the precursors are scavenged is attributed to the existence of spurs, which are characterized by initially higher radical concentrations. The unscavengeable yield of molecular hydrogen, $G_{H_2}^0$, may originate in these "condensed" spurs (see Section V.C).

B. Some Arguments against Intraspur Reactions as a Source of G_{H_2}

Anbar and Meyerstein [133] started from the fact that if there actually are intraspur reactions [Eqs. (28)–(30)], then the isotopic effect must be affected by the pH of the solution. The authors carried out experiments in water in which the ratio of H_2O to D_2O was 1:1. The pH of the solution was varied from 1.1 to 14. The isotopic effect measured on molecular hydrogen was 2.2, irrespective of the pH of the solution. The conclusion is that the origin of primary atomic hydrogen cannot be defined in terms of the diffusion model and intraspur reactions, but is to be sought in the dissociation of excited water molecules. It is worthwhile recalling that the diffusion-kinetic calculation [134] predicts only a slight pH dependence, which could be within the error limits under the above working conditions.

From the effect of NO_3^- ions on the reduction of Ce^{4+} in acid media, Sworski [143] concluded that the precursor of molecular hydrogen, which he suggested to be H_3O, disappears by a first-order process. He proceeds from the fact that concentrations higher than $10^{-2} M$ of NO_3^- effectively inhibit the reduction of ceric ions independently of acidity (0.8 and 0.08 N H_2SO_4) and considers the results to be quantitatively consistent with the assumption that the H_3O intermediate disappears in the spur by a first-order process in competition with its reaction with NO_3^-. The process in question is considered as a pseudomonomolecular process, because the NO_3^- effect is independent of temperature.

Kinetic evidence that "excited water" may be the precursor of primary molecular hydrogen has also been proposed [144]. Here again the effect of nitrate ion concentration on the yield of reduction of ceric ions was studied and found to be consistent with a mechanism in which NO_3^- reacts with electronically excited water molecules. The influence of nitrate ions on the measured yields of molecular hydrogen produced by radiations of different LET was also considered in detail on the basis of the author's meaurements and data from the literature. The absence of a pH influence, as well as the value calculated for the lifetime of e_{aq}^- in the given cases, led the author to the conclusion that the reaction $e_{aq}^- + NO_3^-$ does not occur in spur. He introduces the H_3O species in the reaction scheme and adduces arguments for this from other sources. To explain the effect of LET, he thinks that it is necessary to differentiate intraspur and interspur reactions of intermediates. His interpretations of the importance of "excited water" in water radiolysis are quite diverse. According to one model, the species is identified as the radical pair $(H_3O + OH)$ which is assumed to undergo geminate recombination in a first-order process. In this model, the solute (NO_3^-) would lower G_{H_2} by reacting with H_3O. According to another model, excited water produces the radicals H_3O and OH in a first-order process; these radicals then freely diffuse. The solute reduces G_{H_2} by direct reaction with the "excited water." The dependence of the yield of primary molecular hydrogen on the solute concentration indicates that $\tau_{H_2O^*} = 10^{-9}$ to 10^{-10} sec.

The origin of intraspur molecular hydrogen was considered in the same way, by taking into account the absence of effect of pH on reactions of solutes (nitrates and N_2O) with the precursor of molecular hydrogen [*145*]. When the solution contains a sufficiently high concentration of a hydrated electron scavenger that reacts only slowly with H atoms, then the straight lines representing the dependence of G_{H_2} on the cube root of scavenger concentration should have slopes depending on the pH. They should diminish with decreasing pH because of increasing efficiency of converting e_{aq}^- into H atoms. The authors have shown that there is no pH effect in $NaNO_3$ solutions at pH 1.0, 5.4, and 12.6, and concluded that the origin of G_{H_2} does not lie in e_{aq}^- reactions in the spur. However, as some calculations based on the diffusion-spur model point out [*134*], the pH effect would be negligible in the NO_3^- concentration range used in these experiments.

Faragi and Desalos [*146*] measured molecular hydrogen yields in deaerated solutions of metal ions known as good scavengers for the reducing species. Increasing the concentrations to up to 1 M $NiSO_4$, Tl_2SO_4, $CoSO_4$, and $ZnSO_4$ was found to have no effect on the hydrogen yield, which was observed to be practically constant (about 0.45). Other scavengers [$CdSO_4$, $CuSO_4$, $HgSO_4$, and $Pb(ClO_4)_3$] were found to affect G_{H_2}, but no correlation between the yields measured and the $k_{e_{aq}^-+S}$ was established. These finding led to the conclusion that e_{aq}^- and H are not the precursors of molecular hydrogen, and its origin was attributed to the reaction of H^- with H_2O ($=H_2 + OH^-$). Peled and Czapski [*48*] have considered these results and carried out various experiments. They found that the addition of 10^{-3} M NO_3^- or NO_2^- to studied solutions of Zn^{2+}, Ni^{2+}, and Co^{2+} lowers the

G_{H_2} values in comparison to solutions which contain no nitrate or nitrite ions. It was explained that the role of either NO_3^- or NO_2^- is to react with M^+ (where M is the metal ion) to prevent any other reactions of M^+ which may yield H_2.

C. Intraspur Reactions Are Not the Unique Source of G_{H_2}

In Section V.A we presented arguments for G_{H_2} formation exclusively by the intraspur reactions given by Eqs. (28)–(30). Some arguments against these reactions were given in the preceding section, where different possibilities for the origin of primary molecular hydrogen were also mentioned. The present state of affairs certainly suggests that reactions (28)–(30) are an important but not the unique source of G_{H_2}. It should be noted that quite a long time ago some results [140] pointed to the existence of an easily scavengeable precursor of G_{H_2} and another that was more difficult to scavenge. It is also worth noting a detail in connection with the diffusion-kinetic model and reactions (28)–(30) (see Chapter Six): Good agreement between the theoretical and measured values of $G_{H_2}^0$ is obtained only if it is assumed that there is a small yield of molecular hydrogen ($G_{H_2}^0$) which is produced in the spur, but not by one of these three reactions; according to Schwarz [134], this $G_{H_2}^0$ is approximately 0.175.

D. Direct Observation of the Recombination Reaction of Two e_{aq}^- and the Formation of Molecular Hydrogen

Since the hydrated electron has a characteristic absorption spectrum which allows its disappearance to be followed without difficulty, the recombination reaction of two hydrated electrons should be experimentally observable. Of course, evidence for such a second-order process does not necessarily mean that its product is molecular hydrogen; direct evidence concerning the product would be necessary to verify this point. On this basis it is interesting to consider in some detail the experiments of Dorfman and Taub [55], who were the first not only to show the existence of the reaction of recombination, but also to carry out measurements on the stable products from which they were able to conclude that the product of the reaction is not 2H but H_2. The experimental test for the disappearance of the hydrated electron according to the second-order rate law is presented. By measuring $G(HD)$ and $G(D_2)$ in alkaline solution of C_2H_5OD in D_2O, the authors have shown that this reaction actually gives rise to molecular hydrogen. The mechanism of the radiolysis of aqueous solutions of ethanol is reasonably well established. In alkaline medium, D_2 can arise only directly from the $2e_{aq}^-$ reaction, while

HD can be produced only in the reaction of D atoms with the H atoms from C_2H_5OD. Hence, if in heavy water the reaction of recombination gives no molecular hydrogen but two D atoms, then $G(HD)$ should be high:

$$G(HD) = G_D + G_{e_{aq}^-} > 3.6$$

and $G(D_2) = G_{D_2} \sim 0.5$. If, however, the observed second-order reaction gives only molecular hydrogen directly, then the yield of this product should be high:

$$G(D_2) = G_{D_2} + \tfrac{1}{2} G_{e_{aq}^-} \sim 2 \qquad G(HD) = G_D \sim 0.6$$

The measured values of $G(D_2)$ and $G(HD)$ were 1.96 and 0.63, respectively. Hence, the experimental results are in favor of the assumption that the recombination of two hydrated electrons produces molecular hydrogen. This work also gives the rate constant for the recombination reaction, $2k = 1.3 \times 10^{10}$ M^{-1} sec^{-1}.

The fact that the above experiments confirm the existence of the recombination reaction with formation of H_2 does not prove that this intraspur reaction is the source of G_{H_2}. Such an assumption requires experimental checking of the second-order rate law on the nanosecond (or even lower) scale. Thomas and Bensasson [147] followed the concentration changes of e_{aq}^- during 50 nsec after a 3.5-nsec pulse, and concluded that there is no recombination reaction (see Chapter Six). This negative answer pointed out that the recombination reaction is completed in a time shorter than 1 nsec. However, no evidence for a nonhomogeneous short-lived "spur-type" reaction has been found even on the 10^{-11} sec scale [36].

It is worthwhile noting that recent experiments [148] point out the existence of $(e_2^{-2})_{aq}$, the hydrated electron dimer, assumed to be the intermediate in the reaction leading to the formation of H_2 [Eq. (28)].

References

1. A. H. Samuel and J. L. Magee, Theory of Radiation Chemistry. II. Track Effects in Radiolysis of Water. *J. Chem. Phys.* **21**, 1080 (1953).
2. J. L. Magee, Mechanisms of Energy Degradation and Chemical Change: Effects of Secondary Electrons. *In* "Basic Mechanisms in Radiobiology. II. Physical and Chemical Aspects" (J. L. Magee, M. D. Kamen, and R. L. Platzman, eds.), p. 51, National Research Council, Publ. No. 305, Washington, D.C., 1953.
3. R. L. Platzman, Energy Transfer from Secondary Electrons to Matter. *In* "Basic Mechanisms in Radiobiology. II. Physical and Chemical Aspects" (J. L. Magee, M. D. Kamen and R. L. Platzman, eds.), p. 22. National Research Council, Publ. No. 305, Washington, D.C., 1953.

4. M. J. Day and G. Stein, Chemical Dosimetry of Ionizing Radiations. *Nucleonics* **8**, 34 (1951); G. Stein, Some Aspects of the Radiation Chemistry of Organic Solutes. *Discuss. Faraday Soc.* **12**, 227 (1952).
5. J. H. Baxendale and G. Hughes, The X-Irradiation of Aqueous Methanol Solutions. Part I. Reactions in H_2O. Part II. Reactions in D_2O. *Z. Phys. Chem. (Frankfurt)* **14**, 306 (1958).
6. E. Hayon and J. Weiss, The Action of Ionizing Radiations (200 kV X Rays) on Aqueous Solutions of Mono and Trichloroacetic Acid. *Proc. Int. Conf. Peaceful Uses At. Energy, 2nd Geneva, 1958* **29**, P/1517, 80 (1958).
7. N. F. Barr and A. O. Allen, Hydrogen Atoms in the Radiolysis of Water, *J. Phys. Chem.* **63**, 928 (1959).
8. J. T. Allan and G. Scholes, Effects of pH and the Nature of the Primary Species in the Radiolysis of Aqueous Solutions. *Nature* **187**, 218 (1960).
9. E. Hayon and A. O. Allen, Evidence for Two Kinds of "H atoms" in the Radiation Chemistry of Water. *J. Phys. Chem.* **65**, 2181 (1961).
10. G. Czapski and H. A. Schwarz, The Nature of the Reducing Radical in Water Radiolysis. *J. Phys. Chem.* **66**, 471 (1962).
11. S. Collinson, F. S. Dainton, D. R. Smith, and S. Tazuke, Evidence for the Unit Negative Charge on the "Hydrogen Atom" Formed by the Action of Ionizing Radiation on Aqueous Systems. *Proc. Chem. Soc.* 140 (1962).
12. E. J. Hart and J. W. Boag, Absorption Spectrum of the Hydrated Electron in Water and in Aqueous Solutions. *J. Amer. Chem. Soc.* **84**, 4090 (1962).
13. J. W. Boag and E. J. Hart, Absorption Spectra of "Hydrated" Electron. *Nature* **197**, 45 (1963).
14. J. P. Keene, Optical Absorption in Irradiated Water. *Nature* **197**, 47 (1963).
15. M. Ebert, J. P. Keene, A. J. Swallow, and J. H. Baxendale, eds. "Pulse Radiolysis." Academic Press, New York, 1965.
16. R. F. Gould, ed., "Solvated Electron" (*Advan. Chem. Ser.* **50**). Am. Chem. Soc., Washington, D.C., 1965.
17. H. Greenshields and W. A. Seddon, Pulse Radiolysis. A Comprehensive Bibliography (1960–March 1969), AECL-3524, Chalk River, 1970.
18. E. J. Hart, The Hydrated Electrons. *In* "Actions chimiques et biologiques des radiations" (M. Haissinsky, ed.), Vol. 10, p. 1. Masson, Paris, 1966.
19. D. C. Walker, The Hydrated Electron. *Quart. Rev. Phys. Chem.* **21**, 79 (1967).
20. J. K. Thomas, The Hydrated Electron and the H Atom in the Radiolysis of Water. *In* "Radiation Research, 1966" (G. Silini, ed.), p. 179. North-Holland Publ., Amsterdam, 1967.
21. F. S. Dainton, The Chemistry of the Electron. *In* "Fast Reactions and Primary Processes in Chemical Kinetics" (S. Claesson, ed.), p. 185. Wiley (Interscience), New York, 1967.
22. J. W. Boag, The Events Following Primary Activations. *In* "Radiation Research, 1966" (G. Silini, ed.), p. 43. North-Holland Publ., Amsterdam, 1967.
23. D. C. Walker, Hydrated Electrons in Chemistry. *Advan. Chem. Ser.* **81**, 49 (1968).
24a. M. Anbar, The Reaction of Hydrated Electrons with Inorganic Compounds. *Quart. Rev. Phys. Chem.* **22**, 578 (1968).
24b. M. Anbar, The Reactions of Hydrated Electrons with Organic Compounds. *Advan. Phys. Org. Chem.* **7**, 115 (1969).
25. E. J. Hart, The Hydrated Electron, Its Discovery, Its Spectrum and Its Uses: Introductory Remarks. *In* "Radiation Chemistry of Aqueous Systems" (G. Stein, ed.) p. 73. Wiley (Interscience), New York, 1968.

REFERENCES

26. E. J. Hart, Research Potentials of the Hydrated Electron. *Accounts Chem. Res.* **2**, 161 (1969).
27. P. J. Coyle, F. S. Dainton, and S. R. Logan, The Probable Relaxation Time of the Ionic Atmosphere of the Hydrated Electron. *Proc. Chem. Soc.* 219 (1964).
28. K. H. Schmidt and W. L. Buck, Mobility of the Hydrated Electron. *Science* **151**, 70 (1966).
29. J. H. Baxendale, Redox Potential and Hydration Energy of the Hydrated Electron. *Radiat. Res. Suppl.* **4**, 139 (1964).
30. J. Jortner and R. M. Noyes, Some Thermodynamic Properties of the Hydrated Electron. *J. Phys. Chem.* **70**, 770 (1966).
31. R. M. Noyes, Further Predictions of Thermodynamic Properties of the Hydrated Electron. *Advan. Chem. Ser.* **81**, 65 (1968).
32. J. Jortner, Excess Electron States in Liquids. *In* "Actions chimiques et biologique des radiations" (M. Haissinsky, ed.), Vol. 14, p. 7. Masson, Paris, 1970.
33. J. Rabani, W. A. Mulac, and M. S. Matheson, The Pulse Radiolysis of Aqueous Tetranitromethane. I. Rate Constants and the Extinction Coefficient of e^-_{aq}. II. Oxygenated Solutions. *J. Phys. Chem.* **69**, 53 (1965).
34. E. M. Fielden and E. J. Hart, Primary Radical Yields in Pulse-Irradiated Alkaline Aqueous Solution. *Radiat. Res.* **32**, 564 (1967).
35. S. O. Nielsen, P. Pagsberg, E. J. Hart, H. Christensen, and G. Nilsson, Absorption Spectrum of the Hydrated Electron from 200 to 250 nm. *J. Phys. Chem.* **73**, 3171 (1969).
36. M. J. Bronskill, R. K. Wolf, and J. W. Hunt, Subnanosecond Observations of the Solvated Electron. *J. Phys. Chem.* **73**, 1175 (1969).
37. J. W. Boyle, J. A. Ghormley, C. J. Hochanadel, and J. F. Riley, Production of Hydrated Electrons by Flash Photolysis of Liquid Water with Light in the First Continuum. *J. Phys. Chem.* **73**, 2886 (1969).
38. E. J. Hart, S. Gordon, and E. M. Fielden, Reaction of the Hydrated Electron with Water. *J. Phys. Chem.* **70**, 150 (1966).
39. E. Hart and M. Fielden, Reaction of the Deuterated Electron, e_d^-, with e_d^-, D, OD and D_2O. *J. Phys. Chem.* **72**, 577 (1968).
40. M. Anbar and P. Neta, A Compilation of Specific Bimolecular Rate Constants for the Reactions of Hydrated Electrons, Hydrogen Atoms and Hydroxyl Radicals with Inorganic and Organic Compounds in Aqueous Solutions. *Int. J. Appl. Radiat. Isotopes* **18**, 493 (1967).
41. J. Pukies, W. Roebke, and A. Henglein, Pulsradiolytische Untersuchung einiger Elementarprozesse der Silberreduktion. *Ber. Bunsenges. Phys. Chem.* **72**, 842 (1968).
42. M. Anbar and E. J. Hart, The Reactivity of Metal Ions and Some Oxy Anions Toward Hydrated Electrons. *J. Phys. Chem.* **69**, 973 (1965).
43. G. V. Buxton and F. S. Dainton, The Radiolysis of Aqueous Solutions of Oxybromine Compounds; the Spectra and Reactions of BrO and BrO_2. *Proc. Roy. Soc.* (*London*) **A304**, 427 (1968).
44. B. Čerček, Activation Energies for Reactions of the Hydrated Electron. *Nature* **223**, 491 (1969).
45. E. J. Hart, J. K. Thomas, and S. Gordon, A Review of the Radiation Chemistry of Single-Carbon Compounds and Some Reactions of the Hydrated Electron in Aqueous Solution. *Radiat. Res. Suppl.* **4**, 74 (1964).
46. S. Gordon, E. J. Hart, M. S. Matheson, J. Rabani, and J. K. Thomas, Reactions of the Hydrated Electron. *Discuss. Faraday. Soc.* **36**, 193 (1963).
47. J. K. Thomas, S. Gordon, and E. J. Hart, The Rates of Reaction of the Hydrated Electron in Aqueous Inorganic Solutions. *J. Phys. Chem.* **68**, 1524 (1964).

48. E. Peled and G. Czapski, Studies on the Molecular Hydrogen Formation (G_{H_2}) in the Radiation Chemistry of Aqueous Solutions. *J. Phys. Chem.* **74**, 2903 (1970).
49. M. Anbar and E. J. Hart, On the Reactivity of Hydrated Electrons Toward Inorganic Compounds. *Advan. Chem. Ser.* **81**, 79 (1968).
50. J. H. Baxendale, E. M. Fielden, C. Capellos, J. M. Francis, J. V. Davies, M. Ebert, C. W. Gilbert, J. P. Keene, E. J. Land, A. J. Swallow, and J. M. Nosworthy, Pulse Radiolysis. *Nature* **201**, 468 (1964).
51. O. Mićić and I. G. Draganić, A Study of Some Free-Radical Reactions in Aqueous γ-Radiolysis by Direct Measurements of Cu^+ Intermediate During Irradiation. *J. Phys. Chem.* **70**, 2212 (1966).
52. M. S. Matheson and J. Rabani, Pulse Radiolysis of Aqueous Hydrogen Solutions. I. Rate Constants for Reaction of e_{aq}^- with Itself and Other Transients. II. The Interconvertibility of e_{aq}^- and H. *J. Phys. Chem.* **69**, 1324 (1965).
53. S. Gordon, E. J. Hart, M. S. Matheson, J. Rabani, and J. K. Thomas, Reaction Constants of the Hydrated Electron. *J. Amer. Chem. Soc.* **85**, 1375 (1963).
54. K. Schmidt and E. J. Hart, A Compact Apparatus for Photogeneration of Hydrated Electrons. *Advan. Chem. Ser.* **81**, 267 (1968).
55. L. M. Dorfman and I. A. Taub, Pulse Radiolysis Studies. III. Elementary Reactions in Aqueous Ethanol Solution. *J. Amer. Chem. Soc.* **85**, 2370 (1963).
56. M. Anbar, On the Reactivity of Transition-Metal Complex Ions Toward Hydrated Electrons. *Chem. Commun.* 416 (1966).
57. J. P. Keene, The Absorption Spectrum and Some Reaction Constants of the Hydrated Electron. *Radiat. Res.* **22**, 1 (1964).
58. B. Hickel and K. H. Schmidt, Kinetic Studies with Photogenerated Hydrated Electrons in Aqueous Systems Containing Nitrous Oxide, Hydrogen Peroxide, Methanol, or Ethanol. *J. Phys. Chem.* **74**, 2470 (1970).
59. E. J. Hart and E. M. Fielden, Submicromolar Analysis of Hydrated Electron Scavengers. *Advan. Chem Ser.* **50**, 253 (1965).
60. W. D. Felix, B. L. Gall, and L. M. Dorfman, Pulse Radiolysis Studies. IX. Reactions of the Ozonide Ion in Aqueous Solution. *J. Phys. Chem.* **71**, 384 (1967).
61. J. H. Baxendale, E. M. Fielden and J. P. Keene, The Pulse Radiolysis of Aqueous Solutions of Some Inorganic Compounds, *Proc. Roy. Soc. (London)* **A286**, 320 (1965).
62. G. Meissner and A. Henglein, Die Reaktionen des Schwefelwasserstoffs mit hydratisierten Elektronen und freien Wasserstoffatomen und die Solvatationsenergie des Elektrons im Wasser. *Ber. Bunsenges. Phys. Chem.* **69**, 3 (1965).
63. D. M. Brown and F. S. Dainton, Matrix Isolations of Unstable Lower Valency States of Metal Cations. *Trans. Faraday Soc.* **62**, 1139 (1966).
64. J. H. Baxendale, E. M. Fielden, and J. P. Keene, Absolute Rate Constants for the Reactions of Some Metal Ions with the Hydrated Electron. *Proc. Chem. Soc.* 242 (1963).
65. B. Čerček and M. Ebert, Activation Energies for Reactions of the Hydrated Electron, *J. Phys. Chem.* **72**, 766 (1968).
66. K. D. Asmus and J. H. Fendler, The Reaction of Sulfur Hexafluoride with Hydrated Electrons. *J. Phys. Chem.* **72**, 4285 (1968).
67. M. Anbar, Z. B. Alfassi and H. Bregman-Reisler, Hydrated Electron Reactions in View of their Temperature Dependence. *J. Amer. Chem. Soc.* **89**, 1263 (1967).
68. K. W. Chambers, E. Collinson, F. S. Dainton, W. A. Seddon, and F. Wilkinson, Pulse Radiolysis: Adducts of Vinyl Compounds and Simple Free Radicals. *Trans. Faraday Soc.* **63**, 1699 (1967).

REFERENCES

69. C. L. Greenstock, M. Ng, and J. W. Hunt, Pulse Radiolysis Studies of Reactions of Primary Species in Water with Nucleic Acid Derivatives. *Advan. Chem. Ser.* **81**, 397 (1968).
70. E. J. Hart, S. Gordon, and J. K. Thomas, Rate Constant of Hydrated Electron Reactions with Organic Compounds. *J. Phys. Chem.* **68**, 1271 (1964).
71. B. D. Michael and E. J. Hart, The Rate Constant of Hydrated Electron, Hydrogen Atom, and Hydroxyl Radical Reactions with Benzene, 1,3-cyclohexadiene, 1,4-cyclohexadiene, and Cyclohexene. *J. Phys. Chem.* **74**, 2878 (1970).
72. M. Anbar and E. J. Hart, The Reactivity of Aromatic Compounds Toward Hydrated Electrons. *J. Amer. Chem. Soc.* **86**, 5633 (1964).
73. M. Anbar and E. J. Hart, The Reactions of Haloaliphatic Compounds with Hydrated Electrons. *J. Phys. Chem.* **69**, 271 (1965).
74. R. Braams, Rate Constants of Hydrated Electron Reactions with Amino Acids. *Radiat. Res.* **27**, 319 (1966).
75. A. J. Swallow, Recent Results from Pulse Radiolysis. *Photochem. Photobiol.* **7**, 683 (1968).
76. O. Mićić and I. G. Draganić, Some Reactions of Hydrated Electron in Acid Medium (pH 0.6–4.0). *Int. J. Radiat. Phys. Chem.* **1**, 287 (1969).
77. E. J. Land and M. Ebert, Pulse Radiolysis Studies of Aqueous Phenol. Water Elimination from Dihydroxycyclohexadienyl Radicals to Form Phenoxyl. *Trans. Faraday Soc.* **63**, 1181 (1967).
78. E. J. Hart, E. M. Fielden, and M. Anbar, Reactions of Carbonylic Compounds with Hydrated Electrons. *J. Phys. Chem.* **71**, 3993 (1967).
79. S. E. Benson, "The Foundations of Chemical Kinetics," p. 525. McGraw-Hill, New York, 1960.
80. F. S. Dainton and W. S. Watt, pH Effect in the γ-Radiolysis of Aqueous Solutions. *Proc. Roy. Soc. (London)* **A275**, 447 (1963),
81. A. O. Allen, "The Radiation Chemistry of Water and Aqueous Solutions," p. 75. Van Nostrand-Reinhold, Princeton, New Jersey, 1961.
82. H. Fricke and J. K. Thomas, Pulsed Electron Beam Kinetics. *Radiat. Res. Suppl.* **4**, 35 (1964)
83. F. S. Dainton and S. A. Sills, The Rates of Some Reactions of Hydrogen Atoms in Water at 25°C. *Proc. Chem. Soc. (London)*, 223 (1962).
84. J. P. Sweet and J. K. Thomas, Absolute Rate Constants for H Atom Reactions in Aqueous Solutions. *J. Phys. Chem.* **68**, 1363 (1964).
85. J. P. Keene, Y. Raef, and A. J. Swallow, Pulse Radiolysis Studies of Carboxyl and Related Radicals. *In* "Pulse Radiolysis" (M. Ebert, J. P. Keene, A. J. Swallow, and J. H. Baxendale, eds.), p. 99. Academic Press, New York, 1965.
86. G. Navon and G. Stein, Rate Constants of Some Reactions of H Atoms in Aqueous Solutions. *J. Phys. Chem.* **69**, 1384 (1965).
87. W. G. Rothschild and A. O. Allen, Studies in the Radiolysis of Ferrous Sulfate Solutions. III. Air-Free Solutions at Higher pH. *Radiat. Res.* **8**, 101 (1958).
88. J. H. Baxendale and D. H. Smithies, Die Strahlenchemie in Wasser gelöster organischer Verbindungen. *Z. Phys. Chem. (Frankfurt)* **7**, 242 (1956).
89. J. H. Baxendale and P. L. T. Bevan, Absolute Rates of Hydrogen Atom Reactions in Aqueous Solution. *In* "The Chemistry of Ionization and Excitation" (G. R. A. Johnson and G. Scholes, eds.), p. 253. Taylor and Francis, London, 1967.
90. J. Rabani, On the Reactivity of Hydrogen Atoms in Aqueous Solutions. *J. Phys. Chem.* **66**, 361 (1962).

91. K. D. Asmus, A. Henglein, M. Ebert, and J. P. Keene, Pulsradiolytische Untersuchung schneller Reaktionen von hydratisierten Elektronen, freien Radikalen und Ionen mit Tetranitromethan in wässriger Lösung. *Ber. Bunsenges. Phys. Chem.* **68**, 657 (1964).
92. G. Czapski and G. Stein, The Oxidation of Ferrous Ions in Aqueous Solution by Atomic Hydrogen. *J. Phys. Chem.* **63**, 850 (1959).
93. G. Czapski, J. Jortner, and G. Stein, The Mechanism of Oxidation by Hydrogen Atoms in Aqueous Solution. I. Mass Transfer and Velocity Constants. *J. Phys. Chem.* **65**, 956 (1961).
94. J. Jortner and J. Rabani, The Decomposition of Chloroacetic Acid in Aqueous Solutions by Atomic Hydrogen. I. Comparison with Radiation Chemical Data. *J. Phys. Chem.* **66**, 2078 (1962).
95. G. Navon and G. Stein, The Reactivity of Atomic Hydrogen with Cystein, Cystine, Tryptophan and Tyrosine in Aqueous Solutions. *Israel J. Chem.* **2**, 151 (1964).
96. J. Jortner, M. Ottolenghi, J. Rabani, and G. Stein, Conversion of Solvated Electrons into Hydrogen Atoms in the Photo and Radiation Chemistry of Aqueous Solutions, *J. Chem. Phys.* **37**, 2488 (1962).
97. J. Rabani, The Interconvertibility of e_{aq}^- and H. *Advan. Chem. Ser.* **50**, 242 (1965).
98a. J. Jortner and J. Rabani, The Reactivity of Hydrogen Atoms in Alkaline Solutions. *J. Amer. Chem. Soc.* **83**, 4968 (1961).
98b. J. Jortner and J. Rabani, The Decomposition of Chloroacetic Acid in Aqueous Solutions by Atomic Hydrogen. II. Reaction Mechanisms in Alkaline Solutions. *J. Phys. Chem.* **66**, 2081 (1962).
99. M. Anbar and P. Neta, Reaction of Fluoride Ions with Hydrogen Atoms in Aqueous Solutions. *Trans. Faraday Soc.* **63**, 141 (1967).
100. G. Czapski, J. Jortner, and G. Stein, The Oxidation of Iodide Ions in Aqueous Solution by Atomic Hydrogen. *J. Phys. Chem.* **63**, 1769 (1959).
101. G. Czapski and G. Stein, Oxidation of Ferrous Ion in Aqueous Solution by Atomic Hydrogen, *Nature* **182**, 598 (1958).
102. D. Katakis and A. O. Allen, Mechanism of Radiolytic Oxidation of Ferrous Ion. *J. Phys. Chem.* **68**, 657 (1964).
103a. R. R. Hentz and C. G. Johnson, γ-Radiolysis of Liquids at High Pressures. VII. Oxidation of Iodide Ion by Hydrogen Atoms in Aqueous Solutions. *J. Chem. Phys.* **51**, 1236 (1969).
103b. R. R. Hentz, Farhataziz, D. J. Milner, and M. Burton, γ-Radiolysis of Liquids at High Pressures. I. Aqueous Solutions of Ferrous Sulfate. *J. Chem. Phys.* **46**, 2995 (1967).
104. H. L. Friedman and A. H. Zeltmann, Hydrogen Conversion and Exchange Reactions in Aqueous Solutions Induced by Gamma Rays. *J. Chem. Phys.* **28**, 878 (1958).
105. G. Navon and G. Stein, The Reactivity of Some High- and Low-Spin Iron (III) Complexes with Atomic Hydrogen in Aqueous Solutions. *J. Phys. Chem.* **70**, 3630 (1966).
106. G. Scholes and M. Simić, Reactivity of the Hydrogen Atoms Produced in the Radiolysis of Aqueous Systems. *J. Phys. Chem.* **68**, 1738 (1964).
107. G. Czapski, J. Rabani, and G. Stein, The Reactivity of Hydrogen Atoms with Ethanol and Formate in Aqueous Solutions. *Trans. Faraday Soc.* **58**, 2160 (1962).
108. E. Hayon and M. Moreau, Réactivité des atomes H avec quelques composés organiques et minéraux en solutions aqueuses. *J. Chim. Phys.* **62**, 391 (1965).
109. P. Riesz and E. J. Hart, Absolute Rate Constants for H Atom Reactions in Aqueous Solutions. *J. Phys. Chem.* **63**, 858 (1959).

110. J. H. Baxendale, R. S. Dixon, and D. A. Stott, Reactivity of Hydrogen Atoms with Fe^{3+}, $FeOH^{2+}$ and Cu^{2+} in Aqueous Solutions. *Trans. Faraday Soc.* **64**, 2398 (1968).
111. H. A. Schwarz, Absolute Rate Constants for Some Hydrogen Atom Reactions in Aqueous Solution. *J. Phys. Chem.* **67**, 2827 (1963).
112. J. K. Thomas, The Rate Constants for H Atom Reactions in Aqueous Solutions. *J. Phys. Chem.* **67**, 2593 (1963).
113. J. K. Thomas, Discussion. *Radiat. Res. Suppl.* **4**, 111 (1964).
114. J. Rabani and D. Meyerstein, Pulse Radiolytic Studies of the Competition H + H and H + Ferricyanide. The Absolute Rate Constants. *J. Phys. Chem.* **72**, 1599 (1968).
115. J. H. Baxendale, E. M. Fielden, and J. P. Keene, Pulse Radiolysis of Ag^+ Solutions. *In* "Pulse Radiolysis" (M. Ebert, J. P. Keene, A. J. Swallow, and J. H. Baxendale, eds.), p. 207. Academic Press, New York, 1965.
116. P. Pagsberg, H. Christensen, J. Rabani, G. Nilsson, J. Fenger, and S. O. Nielsen, Far-Ultraviolet Spectra of Hydrogen and Hydroxyl Radicals from Pulse Radiolysis of Aqueous Solutions, Direct Measurement of Rate of H + H. *J. Phys. Chem.* **73**, 1029 (1969).
117. S. Nehari and J. Rabani, The Reaction of H Atoms with OH^- in the Radiation Chemistry of Aqueous Solutions. *J. Phys. Chem.* **67**, 1609 (1963).
118. C. J. Hochanadel, Photolysis of Dilute Hydrogen Peroxide Solution in the Presence of Dissolved Hydrogen and Oxygen, Evidence Relating to the Nature of the Hydroxyl Radical and the Hydrogen Atom Produced in the Radiolysis of Water. *Radiat. Res.* **17**, 286 (1962).
119. J. Halpern and J. Rabani, Reactivity of Hydrogen Atoms toward Some Cobalt (III) Complexes in Aqueous Solutions. *J. Amer. Chem. Soc.* **88**, 699 (1966).
120. M. Haissinsky, Radiolyse γ de solutions alcalines et neutres. II. Effet du pH sur les rendements. *J. Chim. Phys.* **62**, 1149 (1965).
121. G. Czapski and J. Jortner, Role of Ferrous Hydride in the Oxidation of Ferrous Ion by Hydrogen Atoms. *Nature* **188**, 50 (1960).
122. F. S. Dainton and P. Fowles, The Photolysis of Aqueous Systems at 1849 Å. I. Solutions Containing Nitrous Oxide. *Proc. Roy. Soc. (London)* **A287**, 295 (1965).
123. J. Rabani and G. Stein, Yield and Reactivity of Electrons and H Atoms in Irradiated Aqueous Solutions. *J. Chem. Phys.* **37**, 1865 (1962).
124. M. Anbar, D. Meyerstein, and P. Neta, Reactivity of Aromatic Compounds towards Hydrogen Atoms. *Nature* **209**, 1384 (1966).
125. P. Neta and L. M. Dorfman, Pulse-Radiolysis Studies. XIV. Rate Constants for the Reactions of Hydrogen Atoms with Aromatic Compounds in Aqueous Solution. *J. Phys. Chem.* **73**, 413 (1969).
126. S. O. Nielsen, P. Pagsberg, J. Rabani, H. Christensen, and G. Nilsson, Pulse Radiolytic Determination of the Ultraviolet Absorption of Hydrogen Atoms in Aqueous Solutions. *Chem. Commun.* 1523 (1968).
127. H. A. Mahlman, Radiolysis of Nitrous Oxide Saturated Solutions: Effect of Sodium Nitrate, 2-Propanol, and Sodium Formate. *J. Phys. Chem.* **70**, 3983 (1966).
128. C. Lifshitz, Isotope Effects in Neutral H_2O–D_2O Irradiated Solutions and the Nature of the Reducing Radical. *Can. J. Chem.* **40**, 1903 (1962).
129. J. L. Magee, Discussion. *Radiat. Res. Suppl.* **4**, 20 (1964).
130. M. Chouraqui and J. Sutton, Origin of Primary Hydrogen Atom Yield in Radiolysis of Aqueous Solutions. *Trans. Faraday Soc.* **62**, 2111 (1966).
131. A. Appleby, The Hydrogen Atom Yield in Irradiated Aqueous Solutions. *In* "The Chemistry of Ionization and Excitation" (G. R. A. Johnson and G. Scholes, eds.), p. 269. Taylor and Francis, London, 1967.

132. M. Moreau and J. Sutton, Influence of pH on the Yield of Hydrogen Atoms in the Radiolysis of Aqueous Solutions. *Trans. Faraday Soc.* **65**, 380 (1969).
133. M. Anbar and D. Meyerstein, H/D Isotope Effects in the Formation of Hydrogen from the Combination of Two Radicals in Aqueous Solutions. *Trans. Faraday Soc.* **62**, 2121 (1966).
134. H. A. Schwarz, Application of the Spur Diffusion Model to the Radiation Chemistry of Aqueous Solutions. *J. Phys. Chem.* **73**, 1928 (1969).
135. J. Sutton and M. Moreau, On the Origin of the Yield of "Primary Hydrogen Atoms" in Water Radiolysis. *In* "Proceedings of the Second Tihany Symposium on Radiation Chemistry" (J. Dobo and P. Hedvig, eds.), p. 95. Akadémiai Kiado, Budapest, 1967.
136. J. A. Ghormley and C. J. Hochanadel, The Effect of Hydrogen Peroxide and Other Solutes on the Yield of Hydrogen in the Decomposition of Water by γ-Rays. *Radiat. Res.* **3**, 227 (1955).
137. A. R. Anderson and E. J. Hart, Hydrogen Yields in the Radiolysis of Aqueous Hydrogen Peroxide. *J. Phys. Chem.* **65**, 804 (1961).
138. H. A. Schwarz, The Effect of Solutes on the Molecular Yields in the Radiolysis of Aqueous Solutions. *J. Amer. Chem. Soc.* **77**, 4960 (1955).
139. E. Hayon and M. Moreau, Reaction Mechanism Leading to the Formation of Molecular Hydrogen in the Radiation Chemistry of Water. *J. Phys. Chem.* **69**, 4058 (1965).
140. H. A. Mahlman, Hydrogen Formation in the Radiation Chemistry of Water. *J. Chem. Phys.* **32**, 601 (1960).
141. H. A. Mahlman, Activity Concept in Radiation Chemistry. *J. Chem. Phys.* **31**, 993 (1959).
142. H. A. Mahlman, Ceric Reduction and the Radiolytic Hydrogen Yield. *J. Amer. Chem. Soc.* **81**, 3203 (1959).
143. T. J. Sworski, Kinetic Evidence for H_3O as the Precursor of Molecular Hydrogen in the Radiolysis of Water. *J. Amer. Chem Soc.* **86**, 5034 (1964).
144. T. J. Sworski, Kinetic Evidence that "Excited Water" is Precursor of Intraspur H_2 in the Radiolysis of Water. *Advan. Chem. Ser.* **50**, 263 (1965).
145. H. A. Mahlman and T. J. Sworski, On the Nature of the Precursor of Molecular Hydrogen in the Radiolysis of Water. *In* "The Chemistry of Ionization and Excitation" (G. R. A. Johnson and G. Scholes, eds.), p. 259. Taylor and Francis, London, 1967.
146. M. Faraggi and J. Desalos, Effect of Positively Charged Ions on the "Molecular" Hydrogen Yield in the Radiolysis of Aqueous Solutions. *Int. J. Radiat. Phys. Chem.* **1**, 335 (1969).
147. J. K. Thomas and R. V. Bensasson, Direct Observation of Regions of High Ion and Radical Concentration in the Radiolysis of Water and Ethanol. *J. Chem. Phys.* **46**, 4147 (1967).
148. N. Basco, G. A. Kenney, and D. C. Walker, Formation and Photodissociation of Hydrated Electron Dimers. *Chem. Commun.* 917 (1969).

As we noted in Chapter Two, few objections can be made to the assumption that the radical OH and its dimer H_2O_2 are the main, and practically the only oxidizing primary products of water radiolysis. It has been shown that the radical HO_2 also appears quite early in the radiolysis of water as a result of the intraspur reaction between OH and H_2O_2. However, its yield is negligibly low for radiations of low LET (such as, for example, ^{60}Co γ-radiation), but it may be more appreciable in the case of radiolysis induced by heavy particles [1,2]. Nevertheless, in this chapter we shall consider the hydroperoxyl radical because of the considerable role it often plays as a secondary product of the radiolysis of water and aqueous solutions containing oxygen. In this case, the primary reducing products H and e_{aq}^- react very rapidly with oxygen to produce HO_2 and O_2^-, thereby replacing the reactive reducing species by others with mainly oxidizing character.

There is no doubt that over a wide range of pH the hydroxyl radical is a neutral species properly represented by OH. As we shall see, this conclusion follows from experiments in which its reaction rate constant was measured at different ionic strengths. The question arises as to the extent to which its acid form, H_2O^+, is formed in acid media; only a few indications exist in its favor, and there is more evidence to the contrary. In alkaline media the reaction of the hydroxyl radical with OH^- ions leads to the formation of O^- ion radicals. Although this entity is a secondary radical, we shall consider it here at some length; the transformation is very effective in alkaline medium and the behavior of O^- does not seem to be identical in every respect with that of OH, although both are oxidizing species.

Considerations of the origin of the primary molecular product (H_2O_2) shed more light upon the water radiolysis model and hence will be discussed in some detail.

We refer the reader who is interested in the subject treated in this chapter to the review articles [3–5]. An account of the thermochemistry of the radicals OH and HO_2 is given in an article by Gray [6].

CHAPTER FOUR

Primary Products of Water Radiolysis: Oxidizing Species— The Hydroxyl Radical and Hydrogen Peroxide

I. Properties of the Hydroxyl Radical

A. One of the Most Powerful Oxidizing Short-Lived Species

Baxendale [7] gives the value $E^0 = -2.8$ V for the pair OH–OH$^-$ at $[H^+] = 1.0$ M. This means, for example, that OH can oxidize all inorganic ions which can exist in higher valence states (such as Co^{2+}, Ce^{3+}, and Fe^{2+}). Of course, whether the net balance of the reaction observed will always be oxidation depends on what other primary and secondary radicals are in the system (namely, H, e_{aq}^-, and HO_2) and what they will do under given conditions.

The reaction with $KMnO_4$ is, for the present, the only one in which the OH radical is known to behave as a reducing agent.

We have seen in Chapter Three that elucidation of the origin of the primary reducing species has been the object of numerous studies. The origin of the hydroxyl radical lies, with little doubt, in the following reactions:

$$H_2O^+ + H_2O \longrightarrow OH + H_3O^+ \qquad (1)$$

$$\begin{array}{c} H_2O \\ \searrow{\scriptstyle\gamma} \\ \quad H_2O^* \longrightarrow OH + H \qquad (2) \\ \nearrow \\ H_2O^+ + e^- \end{array}$$

These were considered in Sections III and V of Chapter Two.

It has long been known that the photolysis of aqueous solutions of H_2O_2 produces hydroxyl radicals. The mechanism of this reaction is not simple [8]:

$$H_2O_2 + h\nu \longrightarrow 2OH \qquad (3)$$

but under suitably chosen working conditions it can be a very useful source of hydroxyl radicals. Hochanadel [9] reasoned that if the primary oxidizing product of water radiolysis and the OH radical produced in reaction (3) are the same species, then the rate constants of their reactions with different scavengers should be the same within the limits of experimental error. To check this idea experimentally, the author chose the reactions

$$H_2O_2 + OH \longrightarrow H_2O + HO_2 \qquad (4)$$

$$H_2 + OH \longrightarrow H_2O + H \qquad (5)$$

and determined k_5/k_4. The measurements were carried out in a dilute solution of hydrogen peroxide in which H_2 and O_2 were present at various concentrations. The photolysis was brought about by light of a wavelength of 2537 Å. In addition to reactions (3)–(5), the reaction scheme also includes the reactions

$$H_2O_2 + H \longrightarrow H_2O + OH \qquad (6)$$

$$O_2 + H \longrightarrow HO_2 \qquad (7)$$

$$HO_2 + HO_2 \longrightarrow H_2O_2 + O_2 \qquad (8)$$

When experimental conditions were adjusted in such a way that $k_7[O_2] \gg k_6[H_2O_2]$, variation of the ratio of H_2 to H_2O_2 made it possible to determine the ratio k_5/k_4 from the competition diagram. The value obtained was almost the same as that previously measured for the radiolytically produced species in water radiolysis, with the presence of oxygen and hydrogen [10]. On this basis he concluded that in both cases the identical entity is involved.

Classical chemical reactions are also used to produce the OH radical for studies concerning its nature and for comparison with the behavior of radiolytically produced species. In Fenton's reagent, the radical is produced by the reaction of ferrous ions with an excess of H_2O_2:

$$Fe^{2+} + H_2O_2 \longrightarrow Fe^{3+} + OH^- + OH \qquad (9)$$

The same should be true for titanous ions:

$$Ti^{3+} + H_2O_2 \longrightarrow Ti^{4+} + OH^- + OH \qquad (10)$$

It has been reported [11] that relative rate constants for OH radical reactions with different solutes are in quite good agreement when the radicals are produced by Fenton's reagent or by radiations. However, ESR studies do not confirm clearly that OH is the radical produced in reactions (9) and (10).

I. PROPERTIES OF THE HYDROXYL RADICAL

Publications [12–14] even point to the opposite: The active species must be a sort of complex between the reactants rather than an OH radical. The agreement between the relative constants does not necessarily indicate that in reactions (9) and (10) free radicals are directly involved. It is sufficient that the form in which the OH is complexed reacts slowly with the scavenger and is in equilibrium with the free OH [4]. In this connection, the following observation is worth attention: The ESR spectra of the hydroxyl radical generated by the titanous ion and hydrogen peroxide in a flow system show two peaks. The minor peak is attributed to the spectrum of the hydroxyl radical associated with the titanic ion. It decays according to a second-order rate equation. The observed rate constant is about four orders of magnitude lower than that for OH radicals [13].

Some other reactions also lead to the production of OH radicals in radiation chemistry. Thus, the primary reducing species reacting with hydrogen peroxide produces OH radicals as given by Eq. (6) or by

$$H_2O_2 + e^-_{aq} \longrightarrow OH^- + OH \qquad [6a]$$

Nitrous oxide efficiently converts hydrated electrons into hydroxyl radicals:

$$N_2O + e^- \longrightarrow N_2 + O^-(+H_2O = OH + OH^-) \qquad (11)$$

This is certainly one of the most frequently used reactions; in the study of aqueous solutions it was introduced by Proskurnin and Kolotyrkin [15].

B. Absorption Spectrum

Some observations in connection with the absorption of light in water vapor [16] indicated that the OH radical might display a weak absorption in the uv region in liquid water. Identification was not easy, since the molar extinction coefficient is small and many other radicals absorb in the same region [17–19]. Figure 4.1 represents the spectrum of the hydroxyl radical. It is worth noticing that the absorption observed in liquid water is less pronounced than that in the gaseous state; it is also shifted toward shorter wavelengths.

The practical importance of this absorption spectrum is far less than that of the hydrated electron (Chapter Three). The molar extinction coefficient is only $4 \times 10^2 \ M^{-1} \ cm^{-1}$ at 2600 Å [19], and the spectral region in which the absorption appears is not very suitable for routine measurements.

The effect of various scavengers on the intensity of absorption provided evidence that the spectrum can be ascribed to the hydroxyl radical. It was shown that only the substance known as scavenger for hydroxyl radicals decreases the optical density. The conversion of hydrated electrons into hydroxyl radicals increased the measured optical densities. From such experiments the rate constants of some reactions could be calculated in good

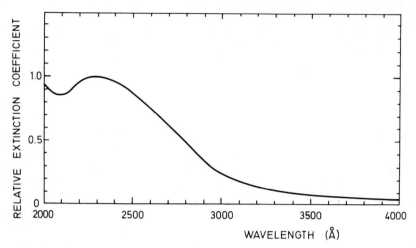

Fig. 4.1. Absorption spectrum of the hydroxyl radical. Normalized to 1.00 at 2300Å. (After Pagsberg et al. [19] and Boyle et al. [18].)

agreement with those of OH radical reactions obtained by indirect measurements.

C. Reactivity

The types of reactions of OH with stable species or free radicals are varied. Electron transfer is the most frequent mechanism of OH-induced oxidation of both inorganic anions and cations. Hydrogen atom abstraction and OH addition are the most common types of reaction with organic molecules. Addition reactions also occur with free radicals.

Electron transfer reactions may give rise to a stable product, as is the case in the well-known reaction of oxidation of the ferrous ion to the ferric ion:

$$Fe^{2+} + OH \longrightarrow Fe^{3+} + OH^- \qquad (12)$$

However, radiation-chemical experiments have also revealed the existence of short-lived products with higher degrees of oxidation; for example, Cu^{3+} [20] and Ag^{2+} [21]. In reactions with anions, electron transfer leads to the formation of a free radical. The reaction with halides has been studied in detail:

$$S^- + OH \longrightarrow S + OH^-$$

where S^- is Cl^-, Br^-, or I^-, and S is the corresponding radical Cl, Br, or I. It was found [22–28] that it is more complex than was previously admitted [29–31]. The halogen atoms react with the solute, $S + S^- \rightarrow S_2^-$, producing ion radicals which also take part in the radiation-induced process.

Abstraction of hydrogen in the reaction of OH with an organic molecule can be written in the general case as

$$\begin{array}{c} R_3 \\ | \\ R_1CH + OH \\ | \\ R_2 \end{array} \longrightarrow \begin{array}{c} R_3 \\ | \\ H_2O + R_1C \\ | \\ R_2 \end{array}$$

where R_1, R_2, and R_3 stand for atomic hydrogen or a certain functional group. These reactions are usually slower than electron transfer. Their rate depends on many factors, particularly on the distribution of electrons in the molecule, which is, of course, affected by the type and arrangement of the functional groups. The disappearance of the organic free radical formed may occur through different reactions. If oxygen is present in the aqueous solution, organic free radicals react with it and the rate constants are often close to the diffusion-controlled value [32]. In the absence of oxygen the disproportionation and dimerization often take place.

The case of the recombination of the α-ethanol radical will be treated at some length as a suitable example of a second-order reaction. Let us first consider the general case where free radical R disappears by the recombination process whence product P is formed:

$$R + R \longrightarrow P$$

$$-\frac{d[R]}{dt} = 2k[R]^2$$

If the radical R has a suitable absorption spectrum in a region where P does not absorb, the reaction may be conveniently followed by observing the variation in [R]. In this case it is easier to deal with optical densities, OD, than with radical concentrations, [R]. According to the Lambert–Beer law,

$$OD = \varepsilon[R]l$$

or

$$[R] = \frac{OD}{\varepsilon l}$$

where ε is the molar extinction coefficient, [R] is the radical concentration, and l is the path of the light in the sample. Introducing $OD/\varepsilon l$ and integrating the above differential expression in the range $t = 0$ to $t = t$, we have

$$\left(\frac{\varepsilon l}{OD}\right)_t - \left(\frac{\varepsilon l}{OD}\right)_0 = 2kt$$

or,

$$\frac{1}{OD_t} = \frac{2k}{\varepsilon l} t + \frac{1}{OD_0}$$

We see that in the plot of reciprocal optical densities against time the slope of the straight line ($\tan \alpha = 2k/\varepsilon l$) enables one to calculate the rate constant

$$2k = \varepsilon l \tan \alpha$$

The value of $2k$ can also be calculated from the initial concentration of the reacting species $[R]_0$ and the experimentally determined initial half-life of the reaction ($t_{1/2}$):

$$2k = \frac{1}{[R]_0 \, t_{1/2}}$$

The intercept on the y-axis gives $[R]_0 = OD_0/\varepsilon l$; hence, the above expression becomes

$$2k = \frac{\varepsilon l}{OD_0 \, t_{1/2}}$$

The α-ethanol radical has a characteristic absorption spectrum with $\varepsilon_{2967} = 240 \ M^{-1} \text{ cm}^{-1}$, and the measurement of OD_{2967} affords a convenient measure of its concentration [33]. A test of the second-order rate law for the disappearance of the α-ethanol radical is shown in Fig. 4.2. We can see that $1/OD_t$ values increase with the time after the pulse. From these data we calculate the slope:

$$\text{ordinate} = 17.1 - 2.9 = 14.2$$
$$\text{abscissa} = 40 \times 10^{-6}$$
$$\tan \alpha = \frac{14.2}{40 \times 10^{-6}} = 0.355 \times 10^6$$

Since $\tan \alpha = 2k/\varepsilon l$, $\varepsilon = 240 \ M^{-1} \text{ cm}^{-1}$, and $l = 16$ cm, $2k = 0.355 \times 10^6 \times \varepsilon \times l = 0.355 \times 240 \times 16 \times 10^6$, that is, $2k = 1.36 \times 10^9 \ M^{-1} \text{ sec}^{-1}$. Another way of calculating $2k$ is to determine the initial half-life of the reaction ($t_{1/2}$). By taking the double initial concentration on the y-axis, we read the corresponding half-life value on the x-axis: $t_{1/2} = 8.1$ μsec. The intersection of the straight line with the y-axis gives $1/OD_0 = 2.9$. Since $2k = \varepsilon l/OD_0 \, t_{1/2}$, substituting $\varepsilon = 240 \ M^{-1} \text{ cm}^{-1}$ and $l = 16$ cm, we obtain

$$2k = \frac{2.9 \times 240 \times 16}{8.1 \times 10^{-6}} = 1.37 \times 10^9 \ M^{-1} \text{ sec}^{-1}$$

Dorfman et al. [34] were the first to show that in aqueous solutions of benzene the addition reaction of the hydroxyl radical occurs. They observed a transient spectrum which was assigned to an OH–benzene adduct (a cyclohexadienyl radical):

$$C_6H_6 + OH \longrightarrow (OH)C_6H_6 \qquad (13)$$

I. PROPERTIES OF THE HYDROXYL RADICAL

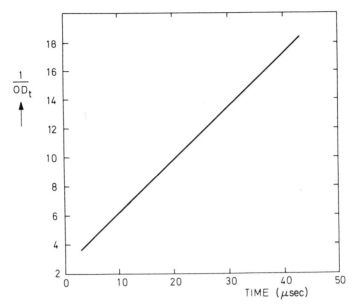

Fig. 4.2. The recombination reaction of the α-ethanol radical in irradiated aqueous solution as an example of a second-order process. (After Dorfman and Taub [33].)

Direct measurements showed that k_{13} is almost diffusion-controlled. This was most surprising, since OH addition must result in the loss of some of the resonance energy of the aromatic nucleus. Nevertheless, subsequent work has confirmed the very large rate constants for reactions of this type [35].

Many absolute rate-constants of OH radical reactions were measured by exploiting the convenient absorption spectrum of the OH adduct of the studied compound. In these experiments it is important to take into account the possibility of simultaneous formation of an H atom adduct; this may have very similar properties and can influence the conclusions regarding the hydroxyl adduct. It has been taken into consideration in the pulse radiolysis of nitrobenzene solutions [36], as well as in pulse studies of aqueous benzoic acid solutions [37].

The fate of an OH adduct has been considered in detail in the case of the cyclohexadienyl radical; Čerček [38] has shown how the substituent X in HOC_6H_5X influences recombination reactions and the reactions with oxygen. Chambers et al. [39–41] have reported data concerning adducts of vinyl compounds.

It is interesting that not only addition reactions to stable molecules but also to free radicals have been observed. It has been shown [42] that the following reaction occurs in aqueous solutions at high electron dose rates obtained with an accelerator:

$$\text{OH} + \text{HO}_2 \longrightarrow \text{H}_2\text{O}_3 \tag{14}$$

In acid medium (0.02 M H_3O^+) the product obtained has a half-life of 2 sec, which decreases with the change of pH. The disappearance of H_2O_3 can be represented as

$$\text{H}_2\text{O}_3 \longrightarrow \text{H}_2\text{O} + \text{O}_2 \tag{15}$$

Polarographic measurements made during irradiation under the same conditions [43], allowing H_2O_2 and O_2 to be followed simultaneously, called in question the importance of reaction (14). Bielski [44] subsequently ascribed to the H_2O_3 species a characteristic absorption spectrum with a peak below 2000 Å. For the molar extinction coefficient he gave values of 100 and 200 M^{-1} cm^{-1} at 2400 and 2000 Å, respectively. The presence of oxygen is necessary for the generation of this product in irradiated water, and scavengers for hydroxyl radicals decrease or completely inhibit its formation.

There is no evidence that the OH radical reacts with oxygen. However, the addition of the O^- ion radical to O_2,

$$\text{O}^- + \text{O}_2 \longrightarrow \text{O}_3^- \tag{16}$$

is not only known to occur, but has also been studied in some detail [45–50]. The product of this reaction—the ozonide ion—has a characteristic absorption maximum at 4300 Å, where the molar extinction coefficient is about 2000 M^{-1} cm^{-1}. The rate constant of reaction (16) is very high, 2.5×10^9 M^{-1} sec^{-1} [45,47]. The half-life of the O_3^- ion radical amounts to several milliseconds and is longer with higher pH [46] and concentration of oxygen in solution [48]. It has been observed to disappear in a first-order process, and also in a second-order process at higher concentrations. The pseudo-first-order process was accounted for as follows [48]: The species which we directly observe by measuring the absorption spectrum of the irradiated solution is not O_3^- but its dimer, which is produced in the very rapid reaction

$$2\text{O}_3^- \rightleftharpoons \text{O}_6^{2-} \tag{17}$$

and which has the same absorption spectrum and molar extinction coefficient as the ozonide ion. Possible reactions proposed for the disappearance of the dimer which satisfy the kinetics of first-order processes were the following:

$$\text{O}_6^{2-} \longrightarrow \text{O}_4^{2-} + \text{O}_2 \tag{18}$$

$$\text{O}_4^{2-} \longrightarrow \begin{cases} 2\text{O}_2^- & (19a) \\ \text{O}_2^{2-} + \text{O}_2 & (19b) \end{cases}$$

In a recent work [50], however, it is assumed that the decay of the ozonide ion in aqueous solution occurs through the thermal dissociation reaction, $\text{O}_3^- = \text{O}_2 + \text{O}^-$, for which the rate constant is 3.3×10^3 sec^{-1} and the activation energy is 11 kcal mole^{-1}.

I. PROPERTIES OF THE HYDROXYL RADICAL

Tables 4.1 and 4.2 summarize the rate constants of OH radical reactions with some free radicals and substances more frequently used in routine experiments.

It should be noted that at the present time our knowledge of OH radical rate constants is much less complete than for the hydrated electron. The lack

TABLE 4.1

RATE CONSTANTS FOR OH AND O^- RADICAL REACTIONS WITH SOME INORGANIC SOLUTES AND FREE RADICALS

Reactant	Rate constant, M^{-1} sec^{-1}	pH	Reference
Br^-	5×10^9	2	[25]
	1.2×10^9	5–9	[26]
CNS^-	2.8×10^{10}	2; Neutral	[28]
	6.6×10^9	Neutral	[51]
$(+O^-)$	1×10^9	13	[52a]
CO_3^{2-}	4.2×10^8	10.76	[53]
$(+O^-)$	$<3 \times 10^6$	13.5	[53]
Cu^{2+}	3.5×10^8	Neutral	[20]
e_{aq}^-	3×10^{10}	11	[54,55]
$(+O^-)$	2.2×10^{10}	13	[54]
$Fe(CN)_6^{4-}$	1.1×10^{10}	3–10	[56]
	1.07×10^{10}	—	[57]
$(+O^-)$	$<7 \times 10^7$	13.5	[57]
H	3.2×10^{10}	—	[58]
	7×10^9	—	[17]
H_2	6×10^7	7	[17]
	4.5×10^7	7	[59]
$(+O^-)$	8×10^7	13	[54]
HCO_3^-	1.5×10^7	8.4	[53]
H_2O_2	4.5×10^7	7	[59]
	1.2×10^7	0.4–3	[58]
	2.25×10^7	—	[60]
	1.7×10^7	Neutral	[61]
$HO_2^-(+O^-)$	7×10^8	13	[49]
	2.74×10^8	13	[62]
I^-	3.4×10^{10}	2; 7	[28]
	1.02×10^{10}	Neutral	[60]
$O_2(+O^-)$	2.5×10^9	13	[45,47]
$O^-(+O^-)$	1×10^9	13	[56,57]
	8×10^8	13	[5]
OH	6×10^9	0.4–3	[56–58]
	5.0×10^9	—	[17,19]
	4×10^9	—	[59]
$(+O^-)$	$\sim 1 \times 10^{10}$	~ 12	[5,57]
OH^-	3.6×10^8		[9,63]

TABLE 4.2

Rate Constants for OH and O⁻ Radical Reactions with Some Organic Solutes

Reactant	Rate constant, $M^{-1} \sec^{-1}$	pH	Reference
Benzene	4.3×10^9	—	[34]
	7.8×10^9	7	[35]
Benzoate ion	6×10^9	6–9.4	[35]
(+O⁻)	$<6 \times 10^6$	13	[50]
Benzoic acid	4.3×10^9	≤ 3	[37]
Benzonitrile	4.9×10^9	7	[35]
Ethanol	1.6×10^9	Neutral	[61]
	1.83×10^9	—	[35]
	1.0×10^9	10.7	[51]
	7.2×10^8	7	[60]
(+O⁻)	8.4×10^8	13	[50]
Formate ion	2.2×10^9	Neutral	[61]
	2.5×10^9	7	[60]
Methanol	9.5×10^8	Neutral	[61]
	8.4×10^8	—	[35]
	4.7×10^8	10.7	[51]
	4.7×10^8	7	[60]
(+O⁻)	5.2×10^8	13	[50]
Nitrobenzene	3.2×10^9	7	[35]
	4.7×10^9	7	[36]
PNDA	1.25×10^{10}	Neutral	[61]
2-Propanol	1.5×10^9	Neutral	[61]
	1.74×10^9	Neutral	[60]

of a suitable absorption spectrum of the OH radical is certainly one important reason. Another reason is that most information comes from competition studies, both from classical and pulsed-beam experiments. A serious inconvenience is not only that they are time-consuming, but often the reaction schemes may be more complex and less understood than admitted. The case of the rate constants for hydroxyl radical reactions with ethanol and methanol is quite illustrative of this. Comparison of published values obtained in pulsed-beam competition experiments shows considerable discrepancies. The values obtained with aromatic compounds [35] as competitors are almost twice as large as those measured with carbonate, thiocyanate, and selenite ions [51] or iodides [60]. Some details will be considered in Section III.

A large number of organic substances [64], in particular nucleic acids [65], have been studied in pulsed-beam experiments in the presence of thiocyanate ion. Although the rate constants reported should be taken with some reserve, they afford a valuable general picture of hydroxyl radical reactivities toward

II. FORMS OF THE HYDROXYL RADICAL

these compounds. In analyzing these and some other data, one finds the following:
- Rate constants for simple aliphatic acids are pH-dependent; the reactivities of the dibasic acids, $HOOC(CH_2)_nCOOH$, increase with increasing values of n.
- For simple straight-chain aliphatic alcohols, rate constants increase with chain length.
- Aliphatic esters are, in general, less reactive than the corresponding alcohols.
- Free amino acids are less reactive than the simple peptides.

II. FORMS OF THE HYDROXYL RADICAL IN IRRADIATED WATER AT VARIOUS pH'S

A. *Kinetic Salt Effect: The Radical Is Uncharged in Neutral Medium*

In Section II of Chapter Three we have seen how the primary kinetic salt effect helped to establish that the reducing species is the hydrated electron, that is, a species carrying unit negative charge. Hummel and Allen [66] have proceeded from the assumption that corresponding experiments with variation of ionic strength of solution and measurement of the rate constant of reactions with the species said to be the hydroxyl radical may give answer as to whether a neutral or a charged particle is in question here. The authors followed formation of H_2O_2 for different ratios of Br^- and C_2H_5OH in oxygenated aqueous solutions. The construction of a competition diagram and calculation of the relative rate constant of reaction of the OH radical with Br^- and ethanol show that ionic strength of solution has no effect on the rate constant. The concentration of the electrolytes $KClO_4$ and $LiClO_4$ present was amounting to as much as 5×10^{-2} M; hence, if a charged particle is in question, the rate constant should have changed appreciably. From these experiments, the unambiguous conclusion follows that the oxidizing radical under the conditions studied (neutral pH) is not charged and is most likely to exist in the form of OH.

B. *Concerning the Existence of $H_2O_{aq}^+$ in Acid Media*

It is certain that the positive water ion, H_2O^+, produced in the physical stage of radiolysis, is not likely to live sufficiently long to react chemically. We have seen (Chapter Two) that its reaction with another water molecule, to which it is bound by hydrogen bonds ($H_2O^+ + H_2O \rightarrow OH + H_3O^+$), takes about 10^{-14} sec. However, some experiments on radiation-induced hydroxylation could be interpreted only by assuming that the hydroxylating agent is not a neutral but a positively charged particle [67]. Its existence was also assumed in explaining the pH dependence of rate constants of hydroxyl radicals with halide ions [68]. These assumptions are not very plausible. It was shown [69] that the relative rate constant of OH reaction in the system iodide–isopropanol is independent of $[H_3O^+]$ at H_2SO_4 concentrations ranging from 0.05 to 1.00 M. This finding points out that the existence of the equilibrium $H_{aq}^+ + OH \rightleftharpoons H_2O_{aq}^+$ also needs experimental verification.

C. Conversion of OH into O^- in Alkaline Media

The dissociation of the hydroxyl radical in alkaline media has been considered frequently. It is represented in terms of formal ionization as

$$OH \rightleftharpoons O^- + H^+ \tag{20}$$

for which $pK_{OH} = 11.9$ is given. This value has been obtained in independent competition experiments with good agreement; Rabani and Matheson [56] were the first to determine it by observing the effect of pH on change in the rate constant of reaction of the hydroxyl radical with ferrocyanide. The oxidation rate decreases with increasing pH and points to alteration of the nature of the OH radical. From the amount of change, the value of pK was derived. Almost the same values were obtained with oxygenated solutions of CNS^- [52a] and deaerated carbonate solutions [53].

As a matter of fact, it seems more convenient to consider the conversion of the hydroxyl radical into O^- simply in terms of the reaction

$$OH + OH^- \longrightarrow O^- + H_2O \tag{21}$$

where the hydroxyl ion is in competition for the OH radical like any other scavenger present; $k_{21} = 3.6 \times 10^8$ M^{-1} sec^{-1} [9,63]. It is easily seen that reaction (21), the conversion of OH into O^-, becomes important in alkaline solutions where the total reactivity of solute S, defined as $k_{OH+S} \times [S]$, is low as compared to $k_{OH+OH^-} \times [OH^-]$. However, a recent study indicates that k_{21} is much larger than the value quoted above. According to these pulsed-beam experiments [52b], an important feature of the chemistry of OH radical in alkaline solution is the equilibrium

$$OH + OH^- \underset{b}{\overset{f}{\rightleftharpoons}} O^- + H_2O \tag{21a}$$

with $k_{21af} = 1.2 \times 10^{10}$ M^{-1} sec^{-1} and $k_{21ab} = 9.2 \times 10^7$ sec^{-1}.

The ion radical O^- absorbs in the uv region, with a peak at 2400 Å and $\varepsilon_{2400} = 240$ M^{-1} cm^{-1} [62]. The mechanism of its decay is not fully understood [5,62]. Like the hydroxyl radical, the O^- ion radical is an oxidizing species, but some differences in their behavior may be quite important:

- The radical ion O^- reacts rapidly with oxygen, whereas the hydroxyl radical is inert or reacts very slowly.
- The radical ion O^- reacts more rapidly with H_2 and H_2O_2 than does the OH radical.
- The recombination of O^- is considerably slower than that of OH.
- In electron transfer reactions, the difference in behavior is great; O^- reacts much more slowly than does OH; CNS^-, CO_3^{2-}, and ferrocyanide are less effectively oxidized (10–10^3 times) by O^-.
- In hydrogen abstraction reactions, O^- and OH seem to be about equally efficient.

III. RELATIVE RATE CONSTANTS OF HYDROXYL RADICAL REACTIONS; COMPETITION KINETICS

In considering reaction rates we have so far dealt mainly with the direct determination of absolute reaction rate constants. These methods are, although in principle simple since the production or disappearance of one reactant is directly followed, rather limited. The fact that insufficient data are available on the characteristic optical and other physicochemical properties of free radicals often makes interpretation of the results arbitrary and sometimes even impossible. On the other hand, the short duration of the processes considered calls for expensive equipment such as pulsed-radiation sources and detection devices of high sensitivity suitable for these measurements.

Another method is more often used. It consists in adding to the irradiated solution two substances which react in competition with the species studied. Observing a reaction product for different ratios of concentrations of these substances, one determines the ratio of the rate constants of the free radical. The accuracy of these relative rate constants is somewhat lower than that of absolute values. The procedure here is also more time-consuming, but its advantage is that the standard equipment for physicochemical analyses suffices.

If the nature of the competition process is simple, the mathematical expressions used in competition kinetics are also simple. Consider the competition of substances A and B in solution for radical R:

$$R + A \longrightarrow P_A$$
$$R + B \longrightarrow P_B$$

where P_A and P_B are the corresponding products. The probability that radical R will react with substance A is given by the expression

$$\frac{k_A[A]}{k_A[A] + k_B[B]}$$

where k_A and k_B are the corresponding rate constants of reactions of the radical R with the substances A and B. If we express this in terms of the radiation-chemical yields of the product formed, $G(P_A)$, and of the reacting radical, G_R, then we have

$$G(P_A) = G_R \frac{k_A[A]}{k_A[A] + k_B[B]}$$

For convenience, in graphically representing experimental data on the yield $G(P_A)$ for different ratios [A]/[B] of substances used as scavengers for radical R, this expression is reduced by suitable transformations to an equation of the type $y = b + ax$:

$$\frac{1}{G(P_A)} = \frac{1}{G_R} \frac{k_A[A] + k_B[B]}{k_A[A]} = \frac{1}{G_R} + \frac{1}{G_R} \frac{k_B[B]}{k_A[A]} \tag{22}$$

We see that here the yield of product [$y = 1/G(P_A)$] is a variable depending on the ratio of concentrations of scavengers ($x = [B]/[A]$). The segment of the ordinate ($b = 1/G_R$) allows one to calculate the radiation-chemical yield of the free radical whose reaction is being studied. From the slope,

$$\tan \alpha = \frac{1}{G_R} \frac{k_B}{k_A}$$

which is calculated from the diagram, and the segment of the ordinate, one calculates the relative rate constant to be determined for the reaction of the radical (k_B/k_A). Measurement of the yield of product P_B may serve in the same way for the determination, if it is more convenient than that of P_A

$$\frac{1}{G(P_B)} = \frac{1}{G_R} + \frac{1}{G_R} \frac{k_A[A]}{k_B[B]} \tag{22a}$$

Figure 4.3 illustrates the method.

It is worthwhile to point out the importance of some difficulties in determinations of the true initial yield, $G^0(P)$, of the stable product which is followed during a competition study. For a reaction product to be measurable it is necessary that it attains a certain minimum concentration required by the sensitivity of the analytical method. This accumulation, however, may have as a consequence the involvement of the product itself in the reaction mechanism, so that its radiation-chemical yield changes. This is most often noticed as a decrease in the yield with increasing absorbed dose. In such cases, the initial yield $G^0(P)$ should be calculated from a diagram in which the measured $G(P)$ values are presented as a function of the dose absorbed. The ordinate for zero abscissa then gives the initial yield $G^0(P)$ to be determined.

III. RELATIVE RATE CONSTANTS   105

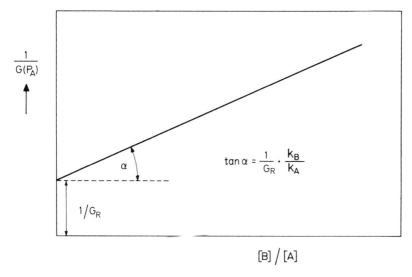

Fig. 4.3. The competition plot used in the relative rate constant determinations.

A cross check is sometimes used for the verification of relative rate constants [70]. For k_A/k_B it consists in the following: The data are also sought for systems B + C and C + D (k_B/k_C and k_C/k_D, respectively). By multiplying $(k_A/k_B)(k_B/k_C)(k_C/k_D)$, the value k_A/k_D is calculated. If the measurements in the system A + D give a value of the relative rate constant in agreement with the one calculated above, then there is the possibility that the verified k_A/k_B ratio is correct. This will not necessarily always be the case; the shortcomings of reaction schemes may influence relative rate constants in such a way that the errors are canceled or diminished and that the agreement is only apparent. It should also be recalled that in the above calculation the errors are accumulated, so that the accuracy of the calculated ratio is quite low.

Most of the published values for $k(OH + S)$ were obtained by the competition procedure described above. Quite a few of them should be accepted with caution, as the interpretations of reaction mechanisms have turned out to be much more complex than was previously assumed. The number of reported rate constants with OH and O^- radicals is still very limited [71]. The reason lies in the difficulty of choosing a competing solute; the mechanism of its change must be well established and the amount of change must be measurable by a simple and accurate method. As we shall see, both conditions are seldom completely satisfied. In the case where benzoic acid was used as the competitor, Mathews and Sangster [72] used a somewhat complex radioactive tracer method; the radioactivity of $^{14}CO_2$, formed by acid decomposition, was measured for various ratios of concentrations of benzoic acid and the solute under study in irradiated solutions. Another procedure was experimentally very simple but, as shown subsequently, the reaction mechanism was more complicated than was initially supposed. This is the case of p-nitrosodimethylaniline (PNDA) which has a suitable absorption spectrum and a large molar extinction coefficient ($\varepsilon = 34200$ M^{-1} cm^{-1} at 4400 Å). The procedure proposed [73a] was extremely simple: The changes of optical density of PNDA should be measured in irradiated solutions containing different ratios of PNDA and the solute studied. It was assumed that the working conditions could be chosen so that the decrease in optical density is the consequence of only OH radical attack on the chromophoric group of PNDA. However, this assumption was subsequently found to be invalid [61,73b] in most of the experiments, since PNDA reacts efficiently not only with OH but with all primary free radicals (OH, H, and e_{aq}^-), as well as with different organic radicals. The reaction scheme is also complicated by the fact that the recombination of the radical formed in the OH + PNDA reaction is very efficient.

In the case of thymine, the experimental procedure seems very simple: The relative rate constant is determined by measuring the decrease of absorption of the chromotropic group on irradiation of aqueous solutions containing different ratios of studied solute and thymine [65,74]. The light-absorbing substance is thymine; it is assumed that only the OH attack causes the decrease of absorption. It is, nevertheless, worth noticing that many of the rate constants obtained by this procedure are close to the values obtained by PNDA, or by pulse competition with CNS$^-$, which we saw to be dubious or even incorrect. It is also known that thymine is destroyed by alcohol radicals even in the presence of oxygen; the attack by H atoms can be very important under certain working conditions, as the rate constant for thymine + H was found to be 8×10^8 M^{-1} sec^{-1} [75]. These findings indicate that a better knowledge of the radiolytic mechanism is still needed if one wishes to avoid surprises similar to those encountered in the PNDA case.

Marketos [76] has shown that the air-saturated aqueous solutions of Safranine T (trade name of 3, 6-diamino-2, 7-dimethyl-10-phenyl-phenazonium chloride) can be used in competition studies at pH 0.4–5.5. It was assumed that in aerated aqueous solutions this substance is irreversibly oxidized by OH radicals only and that H_2O_2, HO_2, and O_2^- do not react with it. Only the rate constants of hydroxyl radical reactions with halide ions were determined, and the system deserves more attention.

In pulsed-beam experiments a simplified competition procedure is quite often used. The measurements consist of optical density determinations of the solution at a fixed time after the electron pulse. In this method, the studied solution contains a substance A which reacts efficiently with the hydroxyl

III. RELATIVE RATE CONSTANTS

radical with known k_{OH+A} to give a product P_A, having a suitable absorption spectrum. Its concentration is then

$$(OD)_0 = [P_A]_0$$

where $(OD)_0$ denotes the absorption measured under strictly defined conditions. If one now adds to the solution a substance B whose rate constant k_{OH+B} is to be determined, then it will scavenge a certain fraction of the OH radicals. As a consequence, the light absorption due to P_A will be decreased, and the new value of the optical density (OD) can be expressed by

$$(OD) = [P_A]_B = \frac{[P_A]_0}{1 + \frac{k_{OH+B}[B]}{k_{OH+A}[A]}}$$

From these relations we obtain the following expression, which is convenient for the construction of the competition diagram:

$$\frac{(OD)_0}{(OD)} = 1 + \frac{k_{OH+B}[B]}{k_{OH+A}[A]} \quad (23)$$

On the x-axis we plot different ratios of concentrations of scavengers [B]/[A], and on the y-axis we plot $(OD)_0/(OD)$. In the case of a true competition, we obtain a competition diagram similar to the one in Fig. 4.3. The rate constant to be determined is calculated as $k_{OH+B} = \tan \alpha \times k_{OH+A}$, where k_{OH+A} is known.

For such pulse-competition experiments the ions CNS$^-$ [51,52a,64,65,77] and I$^-$ [60] were quite often used as scavenger A. Unfortunately, a large number of rate constants of the OH radical obtained by these experiments seem to be practically of no use. Baxendale et al. [28] were the first to show that in the case of CNS$^-$, contrary to what was assumed, the rate constant measured does not concern

$$CNS^- + OH \longrightarrow CNS \quad (24)$$

but $(CNS)_2^-$, formed in the reaction

$$CNS + CNS^- \longrightarrow (CNS)_2^- \quad (25)$$

Also, the reaction scheme may become complicated due to the competition of reaction (25) with some other reactions, where the CNS radical disappears attacking the studied solute and/or some secondary radicals. In the same work [28] the conclusion is also drawn that iodide is not suitable as standard in the competition method of determining OH radical rate constants. Here also, the competition kinetics may not apply because of reactions similar to those mentioned for CNS$^-$ as standard competing solute.

IV. The Hydroperoxyl Radical

A. More Important as a Secondary Radical Than as a Primary Species in the Radiolysis of Aqueous Solutions

The appearance of oxygen in the radiolysis of degassed solutions of ferrous sulfate–copper sulfate led Hart to suppose the existence of the reaction [1]

$$Cu^{2+} + HO_2 \longrightarrow Cu^+ + H^+ + O_2 \qquad (26)$$

Since the solutions were carefully degassed, the author assumed that the origin of the hydroperoxyl radical is in the intraspur reaction

$$H_2O_2 + OH \longrightarrow H_2O + HO_2 \qquad [4]$$

The concentration of radicals in the spur produced by γ rays is low, and it is obvious that G_{HO_2} cannot be large for radiations with low LET values. However, if the assumption is valid, then an increase in LET also leads to an increase in the concentration of hydroperoxyl radicals existing in the beginning of the chemical stage of water radiolysis. Donaldson and Miller [2] found that G_{HO_2} actually increases from 0.026 for ^{60}Co γ-radiation to 0.25 for 5.4-MeV α-radiation. The value of G_{HO_2} for radiations with low LET is small and comparable to experimental error, so that it need not be taken into account in establishing a quantitative reaction scheme in such cases. This is not the case for high-LET studies, but these are still scarce (Chapter Five, Section VI).

However, the hydroperoxyl radical is very important as a secondary radical in the radiolysis of aqueous solutions in the presence of oxygen. The importance lies in the fact that the very rapid reactions (see Chapter Three)

$$O_2 + H \longrightarrow HO_2 \qquad [7]$$

$$O_2 + e_{aq}^- \longrightarrow O_2^- \qquad [7a]$$

both of the order of 10^{10} M^{-1} sec^{-1}, convert reactive species of markedly reducing character into less reactive species of mainly oxidizing character. Since oxygen is usually present in aqueous solutions, the presence of hydroperoxyl radicals is often unavoidable. That is why we shall consider at some length its properties and different reactions, which have been the subject of numerous studies.

B. Properties of the Hydroperoxyl Radical

Baxendale's considerations [7] of redox potentials well illustrate the nature of HO_2. He calculated that in acid solution $E^0 = -1.7$ V for the couple HO_2, H^+/H_2O_2, i.e., for standard change in the free energy of the reaction

$$HO_2 + H^+ + e^- \longrightarrow H_2O_2 \qquad (27)$$

IV. THE HYDROPEROXYL RADICAL

Hence, it follows that under standard conditions the HO_2 radical is as effective an oxidizing agent as, for example, Ce^{4+} ($E^0 = -1.61$ V). Evidence was afforded that it takes part in radiation-induced oxidation of a number of inorganic ions (for example, As^{3+} and Fe^{2+}).

The HO_2 radical may also react as a reducing agent:

$$HO_2 + Ce^{4+} \longrightarrow H^+ + O_2 + Ce^{3+} \qquad (28)$$

The value of E^0 for the couple $H^+, O_2/HO_2$ was calculated to be 0.3 V. In comparison with $E^0 = 2.1$ V for the couple H^+/H the hydroperoxyl radical turns out be a considerably weaker reducing agent.

In aqueous solutions, the reaction

$$HO_2 + R_1 - \underset{\underset{R_2}{|}}{\overset{\overset{R_3}{|}}{C}} H \longrightarrow H_2O_2 + R_1 - \underset{\underset{R_2}{|}}{\overset{\overset{R_3}{|}}{C}} \cdot$$

such as that with ascorbic acid [78], or oxidation of cysteine [79], seldom occurs. The hydroperoxyl radical, like hydrogen peroxide, is most often inert toward organic substances. In view of this fact, its fate lies in dismutation, which leads to the formation and accumulation of H_2O_2 in solution:

$$HO_2 + HO_2 \longrightarrow H_2O_2 + O_2 \qquad [8]$$

The hydroperoxyl radical has a characteristic absorption spectrum with a peak in the uv region at about 2300 Å for HO_2 and about 2500 Å for O_2^-. Their molar extinction coefficients also differ.

A convenient method for producing the hydroperoxyl radical without radiation is the chemical reaction of the ceric ion with hydrogen peroxide. It may be represented by

$$Ce^{4+} + H_2O_2 \longrightarrow Ce^{3+} + HO_2 + H^+ \qquad (29)$$

and Eq. (28). Some reported data point out that the radical chemically produced is identical with the hydroperoxyl formed in irradiated solutions. It has been established [80] that the ESR signal measured under suitable conditions in the course of such a reaction in solution is the same as that obtained by Kroch et al. [81] in irradiated ice. Also, good agreement was found between the values of the rate constant of the recombination reaction, $2HO_2 \rightarrow H_2O_2 + O_2$, by producing HO_2 radicals by chemical means and using the electron beam from an accelerator; the pH of studied solutions varied from 0 to 3 [82]. However, studies show quite clearly that the chemically produced species is not always the free radical (HO_2), but the radical complexed with the ion or molecule which takes part in the chemical process [83–85].

C. The Nature of the Hydroperoxyl Radical in Solutions at Various pH's

The form of the hydroperoxyl radical changes with pH, as shown by kinetic studies of the recombination reaction, $HO_2 + HO_2$ [42,46,82], and the reaction $HO_2 + OH$ [86,87]. The neutral (HO_2) and basic (O_2^-) forms are generally accepted, and pK values of 4.5 [46,86] and 4.8 [88] have been reported for the equilibrium

$$HO_2 \rightleftharpoons H^+ + O_2^- \tag{30}$$

It has also been proposed that in acid medium the protonated form of the hydroperoxyl radical exists,

$$H^+ + HO_2 \rightleftharpoons H_2O_2^+ \tag{31}$$

for which pK values are reported to be 1.0 [82] and 1.2 [86]. Taking into account reactions (30) and (31), the formation of hydrogen peroxide by recombination of hydroperoxyl radicals can be represented by the following reactions:

$$H_2O_2^+ + H_2O_2^+ \longrightarrow H_2O_2 + O_2 + 2H^+ \tag{32}$$

$$H_2O_2^+ + HO_2 \longrightarrow H_2O_2 + O_2 + H^+ \tag{33}$$

$$HO_2 + HO_2 \longrightarrow H_2O_2 + O_2 \tag{8}$$

$$HO_2 + O_2^- \xrightarrow{H_2O} H_2O_2 + O_2 + OH^- \tag{34}$$

$$O_2^- + O_2^- \xrightarrow{2H_2O} H_2O_2 + O_2 + 2OH^- \tag{35}$$

Table 4.3 summarizes the rate constants for the recombination reaction of hydroperoxyl radicals at various pH's. The rate constants obtained at pH values between 0 and 3 are given elsewhere [82]. However, the existence of the $H_2O_2^+$ form is disputed [87] on kinetic grounds. Also, no specific absorption

TABLE 4.3

Rate Constants for HO_2 and O_2^- Radicals

Reaction	Rate constant, M^{-1} sec^{-1}	pH	Reference
$HO_2 + HO_2$	2.2×10^6	2–3	[42]
	2.7×10^6	1.7–3	[46]
	2.5×10^6	2	[47]
	6.7×10^5		[88]
$HO_2 + O_2^-$	5.3×10^7	Neutral	[47]
	7.9×10^7		[88]
$O_2^- + O_2^-$	1.5×10^7	5–8	[42]
	1.7×10^7	5–7	[46]
	$<10^5$		[88]

has been detected in acid solutions which could be attributed to this species. This is surprising, as the effect of ionic strength on the reaction rate constants in acid medium was found to be appreciable and pointed to the existence of a positively charged hydroperoxyl radical form [82]. It is also interesting to note results [88] according to which reaction (35) is practically unimportant because it is very slow; $k_{35} < 10^5 \, M^{-1} \, \text{sec}^{-1}$. This work suggests that the fate of the hydroperoxyl radical is in reactions (8) and (34).

V. Primary Hydrogen Peroxide

The amount of hydrogen peroxide present at the beginning of the chemical stage depends, for a given amount of energy absorbed, only slightly on pH. As we shall see in Chapter Five, $G_{H_2O_2}$ seems to be beyond controversy; it varies from ~ 0.8 for acid medium to about 0.7 for neutral medium, while it may be slightly lower in alkaline solutions. The origin of $G_{H_2O_2}$ also does not seem to present any particular problems. The increase in concentration of any efficient OH radical scavenger leads to a decrease in $G_{H_2O_2}$ pointing to recombination reaction of hydroxyl radicals as the most probable source of primary hydrogen peroxide. This reaction seems to be quite well established.

Nevertheless, the above statement on the origin should be considered merely as a reliable qualitative picture which requires rather extensive quantitative verifications. It is true that the increase in the reactivity toward the OH radical decreases the primary H_2O_2 yields, but there are not many results which confirm that the observed decrease is always proportional to the reactivity, as required by the quantitative free-radical spur diffusion model of water radiolysis. One reason lies in the difficulties in choosing a suitable system: The mechanism of radiolysis should be well established, the rate constant for the reaction of the OH radical with the solute S should be accurately known, and it is especially desirable that $G_{H_2O_2}$ should be directly measured. The difficulties encountered in measuring true initial hydrogen peroxide yields are not negligible. The amount formed by radiation does not remain unchanged; it decomposes in classical chemical reactions with the substances present in the irradiated solutions, or due to attacks by the e_{aq}^- and H species produced by radiation.

In connection with the origin of molecular products (G_M), some other hypotheses have also been put forward, but they are theoretically and experimentally less treated than the above assumption of recombination of radicals [89–91]. We shall, nevertheless, consider one of them, according to which primary hydrogen peroxide is formed in a pseudo-first-order process in which a primary active species reacts with a water molecule directly producing H_2O_2. If this hypothesis, as it appears from the present state of affairs, is far from giving a convincing answer to the question of formation of all $G_{H_2O_2}$, it may nevertheless afford an explanation for the origin of a part of it. This might be important for a better formulation of the quantitative model of water radiolysis.

A. Recombination Reaction of Hydroxyl Radicals

The reactions of recombination of hydroxyl radicals are usually written in such a way as to give hydrogen peroxide as the final product. This also holds for the basic form, the O^- ion radical:

$$OH + OH \longrightarrow H_2O_2 \tag{36}$$

$$OH + O^- \longrightarrow HO_2^- \tag{37}$$

$$O^- + O^- \longrightarrow O_2^{2-} \tag{38}$$

From his experiments, Schwarz [59] calculated $2k_{36} = 8 \times 10^9 \ M^{-1}$ sec^{-1}. The experiments consisted in subjecting water to a pulsed electron beam and studying the steady-state concentration of H_2O_2 as a function of dose rate. A reaction scheme was worked out taking into account the effect of dose rate observed on the measured yield of hydrogen peroxide.

A pulsed electron beam was also used in another work, but molecular hydrogen was measured [58]. The authors varied the pulse length from a time that was short compared to the lifetime of the radical to a time that was relatively long. For this, a special method of treating the data and calculating the results was devised. From the reaction scheme proposed, which involves reaction (36), and experimental data on $G(H_2)$ at pH 3, a value $2k_{36} = 1.2 \times 10^{10} \ M^{-1}$ sec^{-1} was derived. Although this measurement was carried out at pH 3 and in the preceding case in neutral medium, the difference between the two values may seem to be large. However, it should be borne in mind that the reaction rate constant in both cases was determined indirectly, from a complex reaction scheme in which reaction (36) was only one of the constituents.

A more direct measurement was carried out [56,57] by studying reaction (36) in competition with the reaction

$$OH + Fe(CN)_6^{4-} \longrightarrow Fe(CN)_6^{3-} + OH^- \tag{39}$$

the rate constant of which is $k_{39} = 1.1 \times 10^{10} \ M^{-1}$ sec^{-1}. Experimental conditions were arranged for following the recombination reaction. The quantity measured was the yield of the ferricyanide produced, which has $\lambda_{max} = 4200$ Å, with a molar extinction coefficient of 1000 M^{-1} cm^{-1}. Since in this spectral region the hydrated electron also absorbs strongly, various scavengers for e_{aq}^- were added to the system. The value obtained for $2k_{36}$ was $(1.26 \pm 0.16) \times 10^{10} \ M^{-1}$ sec^{-1}. It should be mentioned that this value could not be obtained directly from competition plots, since the reaction scheme was more complex than the one involving only reactions (36) and (39). Hence, by means of computer, experimental curves were correlated with those calculated for three different mechanisms involving reaction sequences which took into account, in addition to the reactions mentioned above, other reactions (H + OH and so on). The value given here was chosen as the most reliable. The authors also carried out measurements in alkaline media at pH 12, 13, and higher than 13. However, the values given for k_{37} and $2k_{38}$ are only approximate ones. Any O_3^- formed may affect the reliability of ferricyanide measurements and hence the information derived from the competition sequences (37) and (39), as well as (38) and (39). Rabani [5] subsequently recalculated these values to be $k_{37} \sim 1 \times 10^{10}\ M^{-1}\ \text{sec}^{-1}$ and $2k_{38} < 1.6 \times 10^9\ M^{-1}\ \text{sec}^{-1}$. It is interesting to note that the recombination of O^- radical ions was found in this work to be considerably slower than that of neutral OH species.

The recombination reaction of OH radicals has been followed directly, and $2k_{36}$ has been calculated from the second-order kinetic plots to be $1.06 \times 10^{10}\ M^{-1}\ \text{sec}^{-1}$ [17,19].

B. Decrease of $G_{H_2O_2}$ with Increasing Reactivity of the OH Scavenger

Sworski [29] was the first to observe that the increase of concentration of bromide ion, a good scavenger for OH radicals, leads to the decrease of primary hydrogen peroxide yields. Other results were subsequently reported, confirming the decrease of measured $G_{H_2O_2}$ with increasing [S]. The scavengers used were Br^- [22,31,92], Cl^- [30], I^- [92], Tl^+ [93–95], Ce^{3+} [93,95], NO_2^- [92,96], $Fe(CN)_6^{3-} + CH_3OH$ [97], V^{4+} [95], H_2O_2 [98], and Cr^{3+} [95]. Schwarz [99] was the first to demonstrate that if $G_{H_2O_2}/G_{H_2O_2}^0$ is plotted against log [S], the resulting series of curves are all of the same shape and can be brought into coincidence by multiplication of the concentrations by an appropriate factor. If the origin of primary hydrogen peroxide lies in the recombination of hydroxyl radicals, as assumed in the quantitative free-radical model of water radiolysis (see Chapter Six), then this normalization factor is in fact the rate constant for the reaction of the OH radical with the solute. However, the comparison of published data does not clearly confirm the basic quantitative assumption, that is, that the $G_{H_2O_2}/G_{H_2O_2}^0$ values depend on the reactivity toward the OH radical only. In the carefully studied case of the halide ions, for example, the efficiency in decreasing the hydrogen peroxide yield does not follow the order of their rate constants with the OH radical. It was also reported [97] that methanol, an efficient scavenger for OH radicals, does not contribute, even at $1M$ concentration, to the $G_{H_2O_2}$ decrease in the system $Fe(CN)_6^{3-} + CH_3OH$ at pH 13. All this may raise doubts that reactions (36)–(38) are the source of $G_{H_2O_2}$. As we shall see, this does not seem to be the case, and the reason for the discrepancy between reported experimental results should be sought mainly in the choice of rate constants, experimental conditions, and in the theoretical interpretation of the results obtained. Often, true initial yields were not measured and the reaction schemes were more complex than was assumed; hence, the calculated values for primary hydrogen peroxide yields were often not true $G_{H_2O_2}$. In a study of the effect of solute concentration on $G_{H_2O_2}$, Draganić and Draganić [100] chose those systems for which the reaction mechanism permits the direct measurement of $G_{H_2O_2}$ by measuring the hydrogen peroxide formed. The solutes used were 1-propanol, ethanol, acrylamide, acetone, and potassium nitrate. The pH of solutions varied between 1.3 and 13. Particular care was taken to ensure that the values derived represent the initial yields. The type and concentration of scavenger were such as to protect H_2O_2 from H or e_{aq}^- attack. When the dosage curves were not straight lines, the corrected values were used as the initial yields. These have been obtained as $G_{H_2O_2}$ readings at zero dose on diagrams where point-by-point peroxide yields were plotted against dose. Figure 4.4 shows the dependence of $G_{H_2O_2}/G_{H_2O_2}^0$ on the reactivity of hydroxyl radical scavenger. Regardless of the nature of OH scavenger and the solution pH, the fractional drop of primary peroxide yield decreases with increasing reactivity as is to be expected if the recombination reactions are the source of

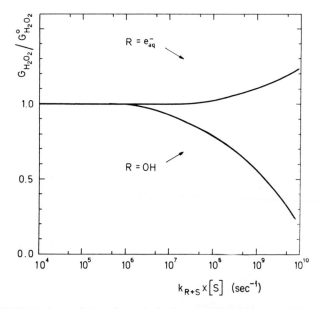

Fig. 4.4. Dependence of the primary hydrogen peroxide yield on reactivity, $k_{R+S} \times$ [S] sec^{-1}. Different scavengers (S) were used at various concentrations (10^{-5}–1 M) and over a wide range of pH's for the solution (1.3–13). (After Draganić and Draganić [100].)

V. PRIMARY HYDROGEN PEROXIDE

primary hydrogen peroxide; $G_{H_2O_2}$ is the primary yield measured in the presence of a scavenger and $G^0_{H_2O_2}$ in the corresponding dilute solution where the solute has no influence on the measured H_2O_2 yield. It has also been shown that the best-fit line through the experimental data agrees well with theoretical [101,102] diffusion kinetic curves.

Another finding from this work may also be of some importance for understanding the origin of primary peroxide. It concerns the measurements of G/G^0 in solutions where the reactivity toward the OH radical is low, but the reactivity toward the hydrated electron increases. As seen in Fig. 4.4, increase in reactivity, contrary to the preceding case, leads here to increase in G/G^0 values. In fact, the quantitative free-radical model predicts such a phenomenon. The increasing reactivity toward e^-_{aq} suppresses the reaction of recombination of primary free radicals:

$$e^-_{aq} + OH \longrightarrow OH^- \tag{40}$$

The remaining hydroxyl radicals disappear partly by reaction (36). Thus, an excess of hydrogen peroxide is produced, larger at higher $k_{e^-_{aq}+S} \times [S]$ values. The trend of experimental values was also in agreement, here, with theoretical [101] diffusion kinetic calculations.

C. Pseudo-First-Order Process as a Possible Source of $G_{H_2O_2}$—Homogeneous Kinetics and the Decrease of $G_{H_2O_2}$ with Increasing Concentrations of Some Scavengers

The doubt that reactions (36)–(38) are the source of primary H_2O_2 led some authors to the conclusion that nonhomogeneous diffusion kinetics cannot quantitatively express the dependence of primary molecular yields on solute concentration [90,91]. An alternative solution consisted in assuming [100]

$$X + H_2O \longrightarrow H_2O_2 \tag{41}$$

$$X + S \longrightarrow P \tag{42}$$

to be the cause of the observed dependence of $G_{H_2O_2}$ on [S]. The nature of the species X is ambiguous; it might be Sworski's "excited water" [90], Voevodski's [89] excited OH radical, or OH^+ according to Anbar [91]. It is essential that use can then be made of the simple competition kinetics expressed by the following equation:

$$\frac{1}{G_{H_2O_2}} = \frac{1}{G^0_{H_2O_2}}\left\{1 + \frac{k_{X+S}[S]}{k_{X+H_2O}[H_2O]}\right\} \tag{43}$$

From the intercept on the competition plot, one should calculate $G^0_{H_2O_2}$, that is the primary peroxide yield measured in dilute solutions when the solute has no effect.

In some cases, the hydrogen peroxide yields measured at various scavenger concentrations satisfy Eq. (43). What is more interesting is that this can be achieved even with some of the experimental data used in constructing the plot in Fig. 4.4, for solutes known as good OH radical scavengers [*100*]. It is interesting to note, however, that $G^0_{H_2O_2}$ values calculated here from the intercepts are lower than the yields measured in corresponding dilute solutions ($<2.5 \times 10^{-3}$ M). This difference is the same for all solutes and at all pH's: $\Delta G^0_{H_2O_2} = 0.12 \pm 0.02$.

We have seen [*90,100*] that the same experimental data can be used to express kinetically two different assumptions concerning the origin of G_M. Nevertheless, before considering the assumption on homogeneous kinetics, it should be stressed that more accurate information is needed concerning the chemical nature of the species called X; furthermore, an explanation is needed of the different mechanisms of scavenging the precursor of the primary H_2O_2, as reflected in the two different $G^0_{H_2O_2}$ values (for dilute and concentrated solutions). Finally, the homogeneous kinetic model should give an explanation of why efficient electron scavengers, such as KNO_3 or acetone, lead to an increase in $G_{H_2O_2}$ with increasing reactivity. Because of these uncertainties, the conclusion is drawn [*100*] that at present the recombination of OH radicals, as given by the diffusion radical model, is the more probable explanation of the primary H_2O_2 yield. Qualitatively, this model predicts the experimentally observed decrease as well as the increase in measured $G_{H_2O_2}$ values at different scavenger concentrations. Quantitatively, the theoretical curves [*101*] satisfy fairly well most of the experimental data. It is important to note that the theoretical curves were obtained using the same parameters as those which furnished good agreement with such a different effect as the variation of yields as a function of LET.

References

1. E. J. Hart, Radiation Chemistry of Aqueous Ferrous Sulfate-Cupric Sulfate Solutions. Effect of γ-Rays. *Radiat. Res.* **2**, 33 (1955).
2. D. M. Donaldson and N. Miller, The Action of α-Particles on Solutions Containing Ferrous and Cupric Ions. *Trans. Faraday Soc.* **52**, 652 (1956).
3. G. E. Adams, The Oxidizing Species in Irradiated Water and Aqueous Solutions. *In* "Radiation Research, 1966" (G. Silini, ed.), p. 195. North-Holland Publ., Amsterdam, 1967.
4. G. Czapski, The Nature of Oxygen-Containing Intermediates in Radiation Chemistry and Photochemistry of Aqueous Solutions. *In* "Radiation Chemistry of Aqueous Systems" (G. Stein, ed.), p. 211. Wiley (Interscience), New York, 1968.
5. J. Rabani, Pulse Radiolysis of Aqueous Alkaline Solutions. *In* "Radiation Chemistry of Aqueous Systems" (G. Stein, ed.), p. 229. Wiley (Interscience), New York, 1968.
6. P. Gray, Chemistry of Free Radicals Containing Oxygen. Part 2. Thermochemistry of the Hydroxyl and Hydroperoxyl Radicals. *Trans. Faraday Soc.* **55**, 408 (1959).

7. J. H. Baxendale, Effects of Oxygen and pH in the Radiation Chemistry of Aqueous Solutions. *Radiat. Res. Suppl.* **4**, 114 (1964).
8. D. E. Lea, The Termination Reaction in the Photolysis of Hydrogen Peroxide in Dilute Aqueous Solutions. *Trans. Faraday Soc.* **45**, 81 (1949).
9. C. J. Hochanadel, Photolysis of Dilute Hydrogen Peroxide Solutions in the Presence of Dissolved Hydrogen and Oxygen. Evidence Relating to the Nature of the Hydroxyl Radical and the Hydrogen Atom Produced in the Radiolysis of Water. *Radiat. Res.* **17**, 286 (1962).
10. C. J. Hochanadel, Effects of Cobalt γ-Radiation on Water and Aqueous Solutions. *J. Phys. Chem.* **56**, 587 (1952).
11. I. Kraljić, Kinetics of OH Radical Reactions in Radiolysis, Photolysis and the Fenton System. *In* "The Chemistry of Ionization and Excitations" (G. R. A. Johnson and G. Scholes, eds.), p. 303. Taylor and Francis, London, 1967.
12. T. Shiga, An Electron Paramagnetic Resonance Study of Alcohol Oxidation by Fenton's Reagent. *J. Phys. Chem.* **69**, 3805 (1965).
13. F. Sicilio, R. E. Florin, and L. A. Wall, Kinetics of the Hydroxyl Radical in Aqueous Solution. *J. Phys. Chem.* **70**, 47 (1966).
14. R. E. Florin, F. Sicilio, and L. A. Wall, The Paramagnetic Species from Titanous Salts and Hydrogen Peroxide. *J. Phys. Chem.* **72**, 3154 (1968).
15. M. A. Proskurnin and V. M. Kolotyrkin, Studies in the Radiation Chemistry of Aqueous Solutions. *Proc. Int. Conf. Peaceful Uses At. Energy, 2nd Geneva, 1958* **29**, P/2022 (1958).
16. O. Oldenberg and F. F. Riecke, Kinetics of OH Radicals as Determined by their Absorption Spectrum. III. A Quantitative Test for Free OH; Probabilities of Transition. *J. Chem. Phys.* **6**, 439 (1938).
17. J. K. Thomas, J. Rabani, M. S. Matheson, E. J. Hart, and S. Gordon, Absorption Spectrum of the Hydroxyl Radical. *J. Phys. Chem.* **70**, 2409 (1966).
18. J. W. Boyle, J. A. Ghormley, C. J. Hochanadel, and J. F. Riley, Production of Hydrated Electrons by Flash Photolysis of Liquid Water with Light in the First Continuum. *J. Phys. Chem.* **73**, 2886 (1969).
19. P. Pagsberg, H. Christensen, J. Rabani, G. Nilsson, J. Fenger, and S. O. Nielsen, Far-Ultraviolet Spectra of Hydrogen and Hydroxyl Radicals from Pulse Radiolysis of Aqueous Solutions. Direct Measurement of the Rate of H + H. *J. Phys. Chem.* **73**, 1029 (1969).
20. J. H. Baxendale, E. M. Fielden, and J. P. Keene, Formation of Cu^{III} in the Radiolysis of Cu^{2+} Solutions. *In* "Pulse Radiolysis" (M. Ebert, J. P. Keene, A. J. Swallow, and J. H. Baxendale, eds.), p. 217. Academic Press, New York, 1965.
21. J. H. Baxendale, E. M. Fielden, and J. P. Keene, Pulse Radiolysis of Ag^+ Solutions. *In* "Pulse Radiolysis" (M. Ebert, J. P. Keene, A. J. Swallow, and J. H. Baxendale, eds.), p. 207. Academic Press, New York, 1965.
22. A. Rafi and H. C. Sutton, Radiolysis of Aerated Solutions of Potassium Bromide. *Trans. Faraday Soc.* **61**, 877 (1965),
23. M. Anbar and J. K. Thomas, Pulse Radiolysis Studies of Aqueous Sodium Chloride Solutions. *J. Phys. Chem.* **68**, 3829 (1964),
24. B. Čerček, M. Ebert, C. W. Gilbert, and A. J. Swallow, Pulse Radiolysis of Aerated Aqueous Potassium Bromide Solutions. *In* "Pulse Radiolysis" (M. Ebert, J. P. Keene, A. J. Swallow, and J. H. Baxendale, eds.), p. 83. Academic Press, New York, 1965.
25. H. C. Sutton, G. E. Adams, J. W. Boag, and B. D. Michael, Radical Yields and Kinetics in the Pulse Radiolysis of Potassium Bromide Solutions. *In* "Pulse Radiolysis" (M. Ebert, J. P. Keene, A. J. Swallow, and J. H. Baxendale, eds.), p. 61. Academic Press, New York, 1965.

26. M. S. Matheson, W. A. Mulac, J. L. Weeks, and J. Rabani, The Pulse Radiolysis of Deaerated Aqueous Bromide Solutions. *J. Phys. Chem.* **70**, 2092 (1966).
27. Farhataziz, Cobalt-60 γ-Radiolysis of Solutions of Potassium Bromide in 0.8 N Sulfuric Acid. *J. Phys. Chem.* **71**, 598 (1967).
28. J. H. Baxendale, P. L. T. Bevan, and D. A. Stott, Pulse Radiolysis of Aqueous Thiocyanate and Iodide Solutions. *Trans. Faraday Soc.* **64**, 2389 (1968).
29. T. J. Sworski, Yields of Hydrogen Peroxide in the Decomposition of Water by Cobalt γ-Radiation. I. Effect of Bromide Ion. *J. Amer. Chem. Soc.* **76**, 4687 (1954).
30. T. J. Sworski, Yields of Hydrogen Peroxide in the Decomposition of Water by Cobalt γ-Radiation. II. Effect of Chloride Ion. *Radiat. Res.* **2**, 26 (1955).
31. A. O. Allen and R. A. Holroyd, Peroxide Yield in the γ-Irradiation of Air-Saturated Water. *J. Amer. Chem. Soc.* **77**, 5852 (1955).
32. G. E. Adams and R. L. Willson, Pulse Radiolysis Studies on the Oxidation of Organic Radicals in Aqueous Solution. *Trans. Faraday Soc.* **65**, 2981 (1969).
33. L. M. Dorfman and I. A. Taub, Pulse Radiolysis Studies. III. Elementary Reactions in Aqueous Ethanol Solution. *J. Amer. Chem. Soc.* **85**, 2370 (1963).
34. L. M. Dorfman, I. A. Taub, and R. E. Bühler, Pulse Radiolysis Studies. I. Transient Spectra and Reaction-Rate Constants in Irradiated Aqueous Solutions of Benzene. *J. Chem. Phys.* **36**, 3051 (1962).
35. P. Neta and L. M. Dorfman, Pulse Radiolysis Studies. XIII. Rate Constants for the Reaction of Hydroxyl Radicals with Aromatic Compounds in Aqueous Solutions, *Advan. Chem. Ser.* **81**, 222 (1968).
36. K. D. Asmus, B. Čerček, M. Ebert, A. Henglein, and A. Wigger, Pulse Radiolysis of Nitrobenzene Solutions. *Trans. Faraday Soc.* **63**, 2435 (1967).
37. R. Wander, P. Neta, and L. M. Dorfman, Pulse Radiolysis Studies. XII. Kinetics and Spectra of the Cyclohexadienyl Radicals in Aqueous Benzoic Acid Solution. *J. Phys. Chem.* **72**, 2946 (1968).
38. B. Čerček, Substituent Effects of Cyclohexadienyl Radicals as Studied by Pulse Radiolysis. *J. Phys. Chem.* **72**, 3832 (1968).
39. K. W. Chambers, E. Collinson, F. S. Dainton, and W. Seddon, Radicals and Radical Anions in Addition Polymerization, Their Spectra and Reactivity. *Chem. Commun.* 498 (1966).
40. K. W. Chambers, E. Collinson, F. S. Dainton, W. A. Seddon, and F. Wilkinson, Pulse Radiolysis: Adducts of Vinyl Compounds and Simple Free Radicals. *Trans. Faraday Soc.* **63**, 1699 (1967).
41. K. W. Chambers, E. Collinson, and F. S. Dainton, Addition of e_{aq}^-, H and OH to Acrylamide in Aqueous Solution and Reactions of the Adducts, *Trans. Faraday Soc.* **66**, 142 (1970).
42. G. Czapski and B. H. J. Bielski, The Formation and Decay of H_2O_3 and HO_2 in Electron-Irradiated Aqueous Solutions. *J. Phys. Chem.* **67**, 2180 (1963).
43. Z. P. Zagorski and K. Sehested, Polarographic Study of Aqueous Oxygenated Solutions Irradiated with 10 MeV Electron Pulses. *In* "Pulse Radiolysis" (M. Ebert, J. P. Keene, A. J. Swallow and J. H. Baxendale, eds.), p. 29. Academic Press, New York, 1965.
44. B. H. J. Bielski. *Proc. Informal Conf. Radiat. Chem. Water, 5th* Univ. of Notre-Dame, Radiat. Lab. AEC Doc. No. COO-38-519, p. 83. Notre Dame, Indiana, 1966.
45. G. E. Adams, J. W. Boag, and B. D. Michael, Rate Constant for Reaction of O^- with Oxygen. *Nature* **205**, 898 (1965).
46. G. Czapski and L. M. Dorfman, Pulse Radiolysis Studies. V. Transient Spectra and Rate Constants in Oxygenated Aqueous Solutions. *J. Phys. Chem.* **68**, 1169 (1964).

REFERENCES

47. G. E. Adams, J. W. Boag, and B. D. Michael, Transient Species Produced in Irradiated Water and Aqueous Solutions Containing Oxygen. *Proc. Roy. Soc. (London)* **A289**, 321 (1966).
48. G. Czapski, Pulse Radiolysis Studies in Oxygenated Alkaline Solutions. *J. Phys. Chem.* **71**, 1683 (1967).
49. W. D. Felix, B. L. Gall, and L. M. Dorfman, Pulse Radiolysis Studies. IX. Reactions of the Ozonide Ion in Aqueous Solutions. *J. Phys. Chem.* **71**, 384 (1967).
50. B. L. Gall and L. M. Dorfman, Pulse Radiolysis Studies. XV. Reactivity of the Oxide Radical Ion and of the Ozonide Ion in Aqueous Solution. *J. Amer. Chem. Soc.* **91**, 2199 (1969).
51. G. E. Adams, J. W. Boag, and B. D. Michael, Reaction of the Hydroxyl Radical. II. The Determination of Absolute Rate Constants. *Trans. Faraday Soc.* **61**, 1417 (1965).
52a. G. E. Adams, J. W. Boag, J. Currant, and B. D. Michael, The Pulse Radiolysis of Aqueous Solutions of Thiocyanate Ion. *In* "Pulse Radiolysis" (M. Ebert, J. P. Keene, A. J. Swallow, and J. H. Baxendale, eds.), p. 117. Academic Press, New York, 1965.
52b. G. V. Buxton, Pulse Radiolysis of Aqueous Solutions. Rate of Reaction of OH with OH^-. *Trans. Faraday Soc.* **66**, 1656 (1970).
53. J. L. Weeks and J. Rabani, The Pulse Radiolysis of Deaereated Aqueous Carbonate Solutions. I. Transient Optical Spectrum and Mechanism. II. pK for OH Radicals. *J. Phys. Chem.* **70**, 2100 (1966).
54. M. S. Matheson and J. Rabani, Pulse Radiolysis of Aqueous Hydrogen Solutions. I. Rate Constants for Reaction of e_{aq}^- with Itself and Other Transients. II. The Interconvertibility of e_{aq}^- and H. *J. Phys. Chem.* **69**, 1324 (1965).
55. S. Gordon, E. J. Hart, M. S. Matheson, J. Rabani, and J. K. Thomas, Reactions of the Hydrated Electron. *Discuss. Faraday Soc.* **36**, 193 (1963).
56. J. Rabani and M. S. Matheson, Pulse Radiolytic Determination of pK for Hydroxyl Ionic Dissociation in Water. *J. Amer. Chem. Soc.* **86**, 3175 (1964).
57. J. Rabani and M. S. Matheson, The Pulse Radiolysis of Aqueous Solutions of Potassium Ferrocyanide. *J. Phys. Chem.* **70**, 761 (1966).
58. H. Fricke and J. K. Thomas, Pulsed Electron Beam Kinetics. *Radiat. Res. Suppl.* **4**, 35 (1964).
59. H. A. Schwarz, A Determination of Some Rate Constants for the Radical Processes in the Radiation Chemistry of Water. *J. Phys. Chem.* **66**, 255 (1962).
60. J. K. Thomas, Rates of Reaction of the Hydroxyl Radical. *Trans. Faraday Soc.* **61**, 702 (1965).
61. J. H. Baxendale and A. A. Khan, The Pulse Radiolysis of *p*-Nitrosodimethylaniline in Aqueous Solution. *Int. J. Radiat. Phys. Chem.* **1**, 11 (1969).
62. J. Rabani, Pulse Radiolysis of Alkaline Solutions. *Advan. Chem. Ser.* **81**, 131 (1968).
63. F. S. Dainton and S. A. Sills, the Rates of Some Reactions of Hydrogen Atoms in Water at 25°C. *Proc. Chem. Soc.* 223 (1962).
64. G. E. Adams, J. W. Boag, J. Currant, and E. D. Michael, Absolute Rate Constants for the Reaction of the Hydroxyl Radical with Organic Compounds. *In* "Pulse Radiolysis" (M. Ebert, J. P. Keene, A. J. Swallow, and J. H. Baxendale, eds.), p. 131. Academic Press, New York, 1965.
65. G. Scholes, P. Shaw, R. L. Willson, and M. Ebert, Pulse Radiolysis Studies of Aqueous Solutions of Nucleic Acid and Related Substances. *In* "Pulse Radiolysis" (M. Ebert, J. P. Keene, A. J. Swallow, and J. H. Baxendale, eds.), p. 151. Academic Press, New York, 1965.
66. A. Hummel and A. O. Allen, Radiation Chemistry of Aqueous Solutions of Ethanol and the Nature of the Oxidizing Radical OH. *Radiat. Res.* **17**, 302 (1962).

67. J. Weiss, Polyfunctional Simple Molecules: Chemical Interest. *Radiat. Res. Suppl.* **4**, 141 (1964).
68. D. Sarrah, Beitrag zur Reaktionsfähigkeit radiolytisch erzeugter OH-Radikale in neutraler und saurer Lösung. *Kernenergie* **8**, 220 (1965).
69. G. Hughes and H. A. Makada, Reactivity of OH and O^- in the Radiolysis of Aqueous Solutions. *Trans. Faraday Soc.* **64**, 3276 (1968).
70. Z. D. Draganić, M. M. Kosanić, and M. T. Nenadović, Competition Studies of the Hydroxyl Radical Reactions in Some γ-Ray Irradiated Aqueous Solutions at Different pH Values. *J. Phys. Chem.* **71**, 2390 (1967).
71. M. Anbar and P. Neta, A Compilation of Specific Bimolecular Rate Constants for the Reactions of Hydrated Electrons, Hydrogen Atoms and Hydroxyl Radicals with Inorganic and Organic Compounds in Aqueous Solutions. *Int. J. Appl. Radiat. Isotopes* **18**, 493 (1967).
72. R. W. Matthews and D. F. Sangster, Measurement by Benzoate Radiolytic Decarboxylation of Relative Rate Constants for Hydroxyl Radical Reactions. *J. Phys. Chem.* **69**, 1938 (1965).
73a. I. Kraljić, and C. N. Trumbore, p-Nitrosodimethylaniline as an OH Radical Scavenger in Radiation Chemistry. *J. Amer. Chem. Soc.* **87**, 2547 (1965).
73b. F. S. Dainton and B. Wiseall, Reactions of Nitrosodimethylaniline with Free Radicals. *Trans. Faraday Soc.* **64**, 694 (1968).
74. G. Scholes and R. L. Willson, γ-Radiolysis of Aqueous Thymine Solutions. Determination of Relative Reaction Rates of OH Radicals. *Trans. Faraday Soc.* **63**, 2983 (1967).
75. H. Loman and J. Blok, On the Radiation Chemistry of Thymine in Aqueous Solution. *Radiat. Res.* **36**, 1 (1968).
76. D. G. Marketos, Rate Constants for Some Reactions of the OH Radical in Irradiated Aqueous Solutions at Different pH Values. *Z. Phys. Chem.* (*Frankfurt*) **65**, 306 (1969).
77. C. L. Greenstock, J. W. Hunt, and M. Ng, Pulse Radiolysis Studies of Uracil and Its Derivatives. *Trans. Faraday Soc.* **65**, 3279 (1969).
78. N. F. Barr and C. G. King, The γ-Ray Induced Oxidation of Ascorbic Acid and Ferrous Ion. *J. Amer. Chem. Soc.* **78**, 303 (1956).
79. A. J. Swallow, The Action of γ-Radiation on Aqueous Solutions of Cysteine. *J. Chem. Soc.* 1334 (1952).
80. E. Saito and B. Bielski, The Electron Paramagnetic Resonance Spectrum of the HO_2 Radical in Aqueous Solutions. *J. Amer. Chem. Soc.* **83**, 4467 (1961).
81. J. Kroh, B. C. Green, and J. W. T. Spinks, Electron Paramagnetic Resonance Studies on the Production of Free Radicals in Hydrogen Peroxide at Liquid Nitrogen Temperature. *J. Amer. Chem. Soc.* **83**, 2201 (1961).
82. B. H. J. Bielski and A. O. Allen, Some Properties of HO_2. *In* "Proceedings of the Second Tihany Symposium on Radiation Chemistry" (J. Dobo and P. Hedvig, eds.), p. 81. Akademiai Kiado, Budapest, 1967.
83. M. S. Bains, J. C. Arthur Jr., and O. Hinojosa, An Electron Spin Resonance Study of Intermediates Formed in Fe^{2+}–H_2O_2 and Ce^{4+}–H_2O_2 Systems in the Presence of Ti^{4+} Ions. *J. Phys. Chem.* **72**, 2250 (1968).
84. G. Czapski, H. Levanon, and A. Samuni, ESR Studies of Uncomplexed and Complexed HO_2 Radical Formed in the Reaction of H_2O_2 with Ce^{4+}, Fe^{2+} and Ti^{3+} Ions. *Israel J. Chem.* **7**, 375 (1969).
85. A. Samuni and G. Czapski, Oxidation of Ce^{3+} by HO_2 Radical and Ce^{3+}–HO_2 Complex Formation. *Israel J. Chem.* **8**, 551 (1970).
86. K. Sehested, O. L. Rasmussen, and H. Fricke, Rate Constants of OH with HO_2, O_2^- and $H_2O_2^+$ from Hydrogen Peroxide Formation in Pulse-Irradiated Oxygenated Water. *J. Phys. Chem.* **72**, 626 (1968).

87. B. H. Bielski and H. A. Schwarz, The Absorption Spectra and Kinetics of Hydrogen Sesquioxide and the Perhydroxyl Radical. *J. Phys. Chem.* **72**, 3836 (1968).
88. J. Rabani and S. O. Nielsen, Absorption Spectrum and Decay Kinetics of O_2^- and HO_2 in Aqueous Solutions by Pulse Radiolysis. *J. Phys. Chem.* **73**, 3736 (1969).
89. V. V. Voevodskii, O mehanizme radioliza vody. *Kinet. Katal.* **2**, 4 (1961).
90. T. J. Sworski, Kinetic Evidence that "Excited Water" is Precursor of Intraspur H_2 in the Radiolysis of Water. *Advan. Chem. Ser.* **50**, 263 (1965).
91. M. Anbar, Water and Aqueous Solutions. In "Fundamental Processes in Radiation Chemistry" (P. Ausloos, ed.), p. 651. Wiley (Interscience), New York, 1968.
92. H. A. Schwarz, *In Proc. Informal Conf. Radiat Chem. Water 5th*, Univ. of Notre-Dame, Radiation Laboratory AEC Doc. No. COO-38-519, p. 51. Notre Dame, Indiana (1966).
93. T. J. Sworski, Mechanism for the Reduction of Ceric Ion by Thallous Ion Induced by Cobalt-60 Gamma Radiation. *Radiat. Res.* **4**, 483 (1956).
94. E. Hayon, Solute Scavenging Effects in Regions of High Radical Concentration Produced in Radiation Chemistry. *Trans. Faraday Soc.* **61**, 723 (1965).
95. J. C. Muller, C. Ferrandini, and J. Pucheault, Vitesses d'oxydation d'ions minéraux par les radicaux OH. III. Vitesse d'oxydation des ions Tl^I et V^{IV}. *J. Chim. Phys.* **63**, 232 (1966).
96. H. A. Schwarz and A. J. Salzman, The Radiation Chemistry of Aqueous Nitrite Solutions: The Hydrogen Peroxide Yield. *Radiat. Res.* **9**, 502 (1958).
97. G. Hughes and C. Willis, Scavenger Studies in the Radiolysis of Aqueous Ferricyanide Solutions at High pH. *Discuss. Faraday Soc.* **36**, 223 (1963).
98. M. Anbar, I. Pecht, and G. Stein, On the Formation of the Molecular Hydrogen Peroxide in the Radiolysis of Aqueous Solutions. *J. Chem. Phys.* **44**, 3635 (1966).
99. H. A. Schwarz, The Effect of Solutes on the Molecular Yields in the Radiolysis of Aqueous Solutions. *J. Amer. Chem. Soc.* **77**, 4960 (1955).
100. Z. D. Draganić and I. G. Draganić, On the Origin of Primary Hydrogen Peroxide Yield in the γ-Radiolysis of Water. *J. Phys. Chem.* **73**, 2571 (1969).
101. A. Kuppermann, Diffusion Model of the Radiation Chemistry of Aqueous Solutions. *In* "Radiation Research, 1966" (G. Silini, ed.), p. 212. North-Holland Publ., Amsterdam, 1967.
102. A. Mozumder and J. L. Magee, A Simplified Approach to Diffusion-Controlled Radical Reactions in the Tracks of Ionizing Radiations. *Radiat. Res.* **28**, 215 (1966).

In the preceding chapters we have seen how the primary products of water radiolysis are produced and what we know about their properties. We have intentionally put aside the question of their radiation-chemical yields, because, owing to its importance as well as its complexity, it calls for a separate consideration.

Different texts can be useful in connection with the subject dealt with in this chapter. The dependence of primary yields on pH has been considered in great detail—in acid and neutral media [1], neutral and alkaline media [2], or over a wide pH range [3, 4]. Some other aspects have also been treated [5–7], and for the effects of dose rate, temperature, and pressure the reader should consult the original literature.

CHAPTER FIVE

Radiation-Chemical Yields of The Primary Products of Water Radiolysis and Their Dependence on Various Factors

I. REMARKS CONCERNING THE RADIATION-CHEMICAL YIELDS OF PRIMARY SPECIES

A. Definition—Experimental Determination

By the radiation-chemical yields of primary species is meant the number of free radicals, G_R (where R is H, OH, or e_{aq}^-), or molecules, G_M (where M is H_2 or H_2O_2), per absorbed 100 eV, present during the chemical stage of water radiolysis. In principle, the determination of radiation-chemical yields reduces to the measurement of the concentration of the radiolytic species per a given amount of absorbed energy. In practice, this is not so simple, because, as we have seen in Chapters Three and Four, the species seldom possess physicochemical properties suitable for direct measurement, such as the absorption spectrum of the hydrated electron. Furthermore, under usual experimental conditions the concentrations of primary species are very low (fractions of a micromole), so that the sensitivity of the technique of measurement becomes a problem. Also, the lifetimes of the primary radical products are very short (of the order of a microsecond), which calls for techniques suitable for very rapid measurements. Moreover, difficulties sometimes arise from the molecular products themselves; even if the most sensitive methods of measurement are used, it is necessary to wait until minimum measurable amounts of the products are accumulated. The accumulation of products very frequently complicates the phenomenon observed, because the product being accumulated may itself be susceptible to radiation-chemical changes.

The most often used method of determining the radiation-chemical yields of primary species consists in the following: Properly chosen solutes, which react with the primary species whose yields are to be determined, are added to the water before irradiation. The chemical changes of the solutes are measured during or after irradiation, and the radiation-chemical yields are then calculated from the amount of these changes. It is necessary to assume a sequence of possible reactions, which is said to be the reaction scheme, or often, although less adequately, the reaction mechanism. The kinetic treatment of a set of equations enables us to relate the amount of observed changes to the amount of studied primary species, i.e., to their radiation-chemical yields. Of course, numerical values obtained in such a way should be taken with reserve, because, as is known in kinetics, more than one reaction scheme may be compatible with one and the same set of values obtained by measurement. Therefore, other quite different chemical systems should be taken into consideration. Moreover, the reaction schemes are to be revised in accordance with up-to-date data on the species involved.

Special attention should be paid to the choice of solute for these experiments. This must satisfy as far as possible the following conditions:

- It should be soluble in water over a wide range of concentrations.
- It should not change chemically over a wide range of pH.
- Its reaction with the primary species studied should be known.
- It should be possible to measure simply and precisely the product of this reaction.
- The product of the reaction should not be susceptible to chemical changes under the working conditions.

Evidently, it is not easy to find a substance which can completely satisfy all the conditions. The difficulties are different. Thus, for example, the use of gases as the scavengers in routine experiments (H_2, O_2, N_2O, etc.) is limited to rather low concentrations. Substantial increases in their concentrations considerably complicate experimental conditions, because increased gas pressure in the reaction vessel is then necessary. Increases in pH may also lead to troublesome complications; they favor the hydrolysis of the salts, decrease their solubility, and often induce the so-called processes of autoxidation, as in the case of sulfite, stannite, or chromite.

Pulse-radiolytic experiments allow G_R to be measured directly. The accuracy of the determination depends mainly on how accurately the absorbed dose and the molar extinction coefficient are known. Only in the case of the hydrated electron have these measurements been of practical interest. As could be seen in Chapters Three and Four, the characteristic absorption spectra and the molar extinction coefficients of the hydrogen atom and the hydroxyl radical are less convenient for such measurements.

B. Effect of Various Factors on G_R and G_M

Since the 1950s, when measurements of the radiation-chemical yields of primary species began to be carried out, several hundred papers reporting data obtained under different experimental conditions have been published. They make it possible to give the following answer to the question as to how many water molecules are decomposed per absorbed 100 eV:

$$3.6 \leqslant G_{-H_2O} \leqslant 4.6$$

This is consistent, with few exceptions, with the results of all related measurements which have been carried out so far, regardless of the system observed (solute, its concentration, the pH of the solution) and of the radiation used to induce water decomposition (LET and dose rate). Although the mean value derived from these measurements ($G_{-H_2O} = 4.1 \pm 0.5$) was satisfactory for many considerations in chemistry, physics, biology, and technology, further studies have been pursued. The accurate knowledge of free-radical yields is especially important for gaining a quantitative picture of water decomposition induced by radiations. It is also important for understanding the mechanism of reactions of these species with different substances as well as for other studies, such as the determination of extinction coefficients and bimolecular rate constants of various intermediates observed in pulsed-beam experiments.

In considering various factors which might affect the radiation-chemical yields of radical (G_R) and molecular (G_M) products, it should be borne in mind that these are produced in very fast processes; hence, they can be affected by only a limited number of factors. Such factors are in the first instance those which affect the spatial distribution of primary species, the LET of radiation (in thousand electron volts per micron) and the dose rate (in electron volts per milliliter per second). The larger the amount of energy deposited in a given volume and at a given instant of time, that is, the higher the LET or the dose rate, the closer is the approach of ionized and excited water molecules and thereby also of species originating from them. This also indicates a higher probability of occurrence of certain reactions which are not manifested at low LET and dose rates.

Another important factor in deriving the values of primary yields from experimental measurements is the reactivity of the substance used as the scavenger. The reactivity is defined here as $k_{R+S} \times [S]$ (in reciprocal seconds), where k_{R+S} (in liters per mole per second) is the rate constant for the reaction of the primary radical with the solute whose concentration S is expressed in moles. When the reactivity is sufficiently high, the reaction solute radical may enter in competition with the recombination reaction $(R + R \to M)$ and thereby influence our conclusions on G_R and G_M.

The concentration and nature of solute are known to have an effect on the structure of water [8–10]. No data are available on whether and how these structural changes may affect the radiation-chemical yields of ionized and excited species and thereby G_R and G_M. In this connection, it should be pointed out that the increase of temperature and pressure, which also leads to a change in the ratio of associated to monomeric water, was found to have no significant effect on the yields of primary species.

So far, relatively little attention has been paid to measurements in heavy water. A longer relaxation time would be indicative of somewhat increased yields of free radicals and decreased yields of molecular products with respect to ordinary water. Available experimental data seem to support this assumption.

C. Equation of Material Balance for Water Decomposition

In studying radiation-chemical yields, use is widely made of the assumption of material balance which was first formulated by Allen [11,12]. If we examine the oxidizing products of water radiolysis, we see that for the production of an OH radical one water molecule must be decomposed, whereas for the production of a H_2O_2 molecule two water molecules must be decomposed. Hence,

$$G_{-H_2O} = G_{OH} + 2G_{H_2O_2} \quad (1)$$

Similarly, for reducing products we have

$$G_{-H_2O} = G_H + G_{e_{aq}^-} + 2G_{H_2} \quad (2)$$

Use is more often made of the equation derived from the preceding two:

$$G_{OH} + 2G_{H_2O_2} = G_H + G_{e_{aq}^-} + 2G_{H_2} \quad (3)$$

It relates the radiation-chemical yields of oxidizing species to those of reducing species. Its practical significance lies in the fact that it simplifies the experimental approach; it suffices to determine three of the four unknown quantities, G_{OH}, G_{red} $(= G_H + G_{e_{aq}^-})$, $G_{H_2O_2}$ and G_{H_2}, to calculate the fourth.

In Eqs. (1) and (2), G_{-H_2O} is understood to be the number of water molecules decomposed during the chemical stage of radiolysis. For the sake of preciseness, it should be recalled that the total number of decomposed water molecules is large. In addition to the results of measurement expressed by Eqs. (1) and (2), the yield of total water decomposition must contain a correction factor for water reformation. This last, roughly speaking, is equal to the sum of the yields of molecular products. Hochanadel [13] proposes

$$G^{tot}_{-H_2O} = G^{net}_{-H_2O} + n(G_{H_2} + G_{H_2O_2}) \qquad (4)$$

where $n = 1$ for radiations with high LET (charged particles), and $n = 1.2$ for radiations with low LET (^{60}Co γ-rays).

In Chapter Four we have seen that, in certain cases, HO_2 must also be taken into account as a primary product of water radiolysis. Equation (3) then assumes the form

$$G_{OH} + 2G_{H_2O_2} + 3G_{HO_2} = G_H + G_{e^-_{aq}} + 2G_{H_2} \qquad [3a]$$

II. EFFECT OF SCAVENGER REACTIVITY

If a substance S reacts with a short-lived species R which is a primary product of water radiolysis, to give the product P according to the reaction

$$R + S \longrightarrow P$$

then the increase of the concentration of S should be followed by the increase of $G(P)$. This increasing process should continue until all the species R escaping the recombination (R + R = M) are scavenged. The effectiveness with which the substance S reacts with the species R also depends on the reaction rate constant k_{R+S}. Figure 5.1 shows how the reactivity ($k_{R+S} \times [S]$, in reciprocal seconds) toward OH radicals increases $G(CO_2)$ in oxygenated solutions of sodium formate [14,15]; it also presents the increase of $G(N_2)$ with increasing reactivity toward e^-_{aq} in aqueous solutions of N_2O in the presence of ethanol [16] and nitrite [17].

If we examine these data, we can easily understand why the scavenger concentration in the solution studied is important if $G(P)$ is to be taken as a measure of G_R. Many contradictions in conclusions on primary yields were due to the fact that measurements of stable products were carried out in a narrow concentration range. Sometimes these concentrations were insufficient and the G_R values derived were low, or inversely.

Qualitatively, the curves of Fig. 5.1 can be explained in a simple way. At very low concentrations, the scavenger cannot react with all radicals because their other reactions with molecular products or impurities are also effective; at moderately low concentrations it removes all those R's which are in the bulk of the solution. When the reactivity is sufficiently high, the scavenger S also competes for the species R from the intraspur reactions. Then increase

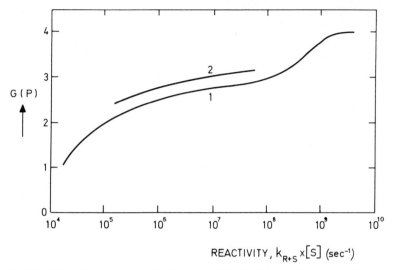

Fig. 5.1. Effect of scavenger reactivity on the yield of stable products. Curve 1 represents the reactivity toward e_{aq}^- and refers to $G(N_2)$ formed in aqueous solutions of N_2O [16, 17]. Curve 2 represents the reactivity toward OH radicals and refers to $G(CO_2)$ measured in $HCOONa + O_2$ [14, 15].

in $G(P)$ begins to occur on account of G_M. Sworski was the first to show a decrease in $G_{H_2O_2}$ due to the increase in concentration of halide ions [18,19]. Schwarz [20] was the first to report a number of experimental data from which it was evident that the increase in concentration of scavenger for short-lived reducing species leads to a decrease in G_{H_2}. Other authors also confirmed that G_M decreases with increasing concentration of scavenger for the precursor of the primary molecular products [21–35]. It has also been shown [3,36] that many discrepancies in G_R values obtained by different authors can be removed if the reactivities are taken into account.

In the next section we shall present a method of calculating the corrections for the increased reactivities [37]. Here, only the empirical procedures are described; they avoid some uncertainties involved in the calculations. In the early days of radiation chemistry it was customary to present $G(P)$ values in a diagram as a function of scavenger concentration. The plateau value was taken as the "correct" value. A disadvantage of such a procedure is that the plateau may appear in the region of low reactivity and give a low value for the "correct" product yield. The reason for this is the loss of primary radicals in a competition between the solute and some impurities, or radiolytic products, which is otherwise absent. Some authors [38,39] represent $G(P)$ values as a function of concentration given as $4 + \log [S]$. The value of the ordinate for zero abscissa is taken as the correct yield. As can be seen,

II. EFFECT OF SCAVENGER REACTIVITY

such an extrapolation gives a yield which would be measured in a 10^{-4} M solution, that is, in a dilute solution where the intraspur reactions are not markedly affected and $G(P)$ constitutes a measure of free radicals in the bulk of the solution only. The objection may be raised that, as we have seen, the concentration does not have the same effect as the reactivity. Also, such an extrapolation gives more weight to measurements in dilute solutions which are less reliable due to reactions with impurities, or other secondary reactions, which the reaction scheme proposed usually leaves out of account.

In the case where the concentration of S is such that a decrease in G_M occurs, the correction was most often made by representing measured G_M values as a function of $[S]^{1/3}$ [18,19,40]. The ordinate for zero abscissa gives the corrected G_{H_2} or $G_{H_2O_2}$. These values are in quite good agreement with those measured in dilute solutions. In certain cases [41,42], similar corrections were made by representing $G(P)$ as a function of $[S]^{1/3}$, but the physical meaning of such a mode of representation is not clear.

From Fig. 5.1 we see that the purpose of the correction should be to take into account the contribution of intraspur reactions. In the ideal case, the corrected $G(P)$ represented as a function of $[S]$ should give only one plateau value. This corresponds to what should be measured in dilute solution in the absence of side reactions. As we shall see in Chapter Six, the quantitative model of water radiolysis is not sufficiently developed and the calculation does not always allow one to get a real plateau value (Section IIIB), especially if the measurements are performed over a wide concentration range (for example, 10^{-6} M to a few moles per liter). But the reason does not lie only in the calculation. In dilute solutions the measurements are quite often unreliable. Low scavenger concentrations require low absorbed doses (it is desirable that these are even lower than 1 krad) in order to avoid appreciable destruction of the solute. In this case the impurities in water and chemicals may play an important role, which is not always easy to control. Furthermore, a suitable sensitive, and accurate analytical method is often lacking; the lower limit of sensitivity of the analytical methods used in routine practice is already attained at doses in the range 10–1 krad. In concentrated solutions the measurements generally raise no difficulties, but the direct action of radiation on solute, as well as the interference of solutes in the earlier stages of radiolysis, might introduce uncertainties in the reaction scheme. Under this condition the correction should not take into account only the contribution of intraspur reactions, but also the product arising from the effects mentioned above. For the present, such effects are still not well established, and the corrections are very unreliable. A recent work on concentrated nitrate solutions under various working conditions [43,44] gives very useful information concerning the contributions of direct and indirect effects.

III. The G_R and G_M Values for γ-Irradiated Water at Neutral pH

A. Mechanism of the Radiolysis of HCOONa + O_2—Use of Steady-State Kinetics for Calculation of Primary Yields

When aqueous solutions of formic acid and oxygen are subjected to the action of ^{60}Co γ-radiation, the stable radiolytic products CO_2, H_2O_2, and H_2 are produced in amounts proportional to the absorbed dose [14,15]. Under properly chosen experimental conditions, the reducing species are efficiently scavenged by molecular oxygen:

$$O_2 + H \longrightarrow HO_2 \qquad k_5 = 1.9 \times 10^{10} \ M^{-1} \ \text{sec}^{-1} \qquad (5)$$

$$O_2 + e_{aq}^- \longrightarrow O_2^- + H_2O \qquad k_6 = 1.88 \times 10^{10} \ M^{-1} \ \text{sec}^{-1} \qquad (6)$$

(Unless otherwise specified, the reaction rate constants quoted in this chapter are taken from the works considered and can be found in corresponding references.)

The formate ion does not compete with oxygen for hydrated electrons as

$$k_{e_{aq}^- + HCOO^-} < 10^6 \ M^{-1} \ \text{sec}^{-1}.$$

However, the reaction

$$HCOO^- + H \longrightarrow H_2 + COO^- \qquad k_7 = 2.5 \times 10^8 \ M^{-1} \ \text{sec}^{-1} \qquad (7)$$

should be taken into account for $[HCOO^-] \geqslant 1 \times 10^{-3} \ M$ as, under the above experimental conditions, $[O_2] = 1 \times 10^{-3} \ M$. The formate ion reacts with the hydroxyl radical:

$$HCOO^- + OH \longrightarrow OH^- + HCOO \ (\text{or} \ H_2O + COO^-)$$
$$k_8 = 2.5 \times 10^9 \ M^{-1} \ \text{sec}^{-1} \qquad (8)$$

The HCOO (or COO^-) radicals produced react effectively with oxygen [45]:

$$O_2 + HCOO \ (\text{or} \ COO^-) \longrightarrow HO_2 \ (\text{or} \ O_2^-) + CO_2$$
$$k_9 = 2.4 \times 10^9 \ M^{-1} \ \text{sec}^{-1} \qquad (9)$$

This reaction accounts for high values of measured H_2O_2 and CO_2 yields. The carboxyl radical reacts with formic acid and hydrogen peroxide, but not in the presence of oxygen. This was shown by the absence of CO as a radiolytic product [46], also confirmed in another work [14], and by the inhibitory action of oxygen on the chain reaction in the $HCOOH + H_2O_2$ system [47]. Furthermore, it is known that carboxyl ion radicals dimerize efficiently [48] ($2k = 1 \times 10^9 \ M^{-1} \ \text{sec}^{-1}$) to give oxalic acid. However, measurements using ^{14}C-labeled formate [49] showed that no oxalic acid is formed in the presence of oxygen. Hence, it follows that the radicals produced in reactions (7) and (8) disappear by reaction (9) only.

III. THE G_R AND G_M VALUES

The hydroperoxyl radicals formed in reactions (5), (6), and (9) undergo disproportionation,

$$2HO_2 \text{ (or } 2O_2^-) \longrightarrow H_2O_2 + O_2 \tag{10}$$

where $k_{HO_2+HO_2} = 2.7 \times 10^6 \ M^{-1} \ sec^{-1}$ and $k_{O_2^-+O_2^-} = 1.7 \times 10^7 \ M^{-1} \ sec^{-1}$, according to Czapski and Dorfman [50]. In another work, these rate constants are reported to be $6.8 \times 10^5 \ M^{-1} \ sec^{-1}$ and $<10^5 \ M^{-1} \ sec^{-1}$, respectively [51]. The fate of the hydroperoxyl radicals is also assumed to be in the $HO_2 + O_2^-$ reaction, with a rate constant of $7.9 \times 10^7 \ M^{-1} \ sec^{-1}$. Reaction (10) accounts for the high yields of H_2O_2 in the irradiated solutions. It should be pointed out that reaction (5) effectively protects H_2O_2 from H atom attack, as $k_{H+H_2O_2} = 5 \times 10^7 \ M^{-1} \ sec^{-1}$ and $[H_2O_2] \ll [O_2]$. This is not quite the case with reaction (6). Experimental curves show that an increase in the absorbed dose, that is, a decrease in O_2 concentration and accumulation of H_2O_2, leads to a slight decomposition due to e_{aq}^- attack. However, even in the most inconvenient case, the correction does not exceed 5% and does not essentially affect the conclusions.

The concentration of H_2 in the solutions is lower than $10^{-5} \ M$, which, since $k_{OH+H_2} = 4.5 \times 10^7 \ M^{-1} \ sec^{-1}$, rules out a competition with reaction (8). The concentration of CO_2 or HCO_3^- in irradiated solutions is considerable (about $2 \times 10^{-5} \ M$). Some fast reactions of these species with the primary products of water radiolysis are known [48,52,53]. Still, as can be seen, their presence does not essentially affect the scheme given by reactions (5)–(10). The reaction

$$CO_2 + e_{aq}^- = COO^- + H_2O \quad k = 7.7 \times 10^9 \ M^{-1} sec^{-1},$$

in the most inconvenient case, can consume 4% of hydrated electrons in competition with reaction (6). Since the fate of the COO^- formed is in reaction (9), it still holds stoichiometrically that only reaction (6) takes place. Bicarbonate ion reacts with OH radicals ($k = 1 \times 10^7 \ M^{-1} \ sec^{-1}$), but its concentration is insufficient to compete with reaction (8) for OH radicals. The reactions of HCO_3^- with hydrogen atoms or hydrated electrons are slow (less than $10^6 \ M^{-1} \ sec^{-1}$), as are the CO_2 reactions with H or OH radicals. Hence, it follows that the stable radiolytic products in the system $HCOONa + O_2$, under properly chosen conditions, cannot affect the simple reaction mechanism even in the case where the irradiation conditions are not strictly initial. If we note that in the kinetic treatment of the data obtained in solutions where $[O_2] = 1 \times 10^{-3} \ M$ and $[HCOO^-] < 1 \times 10^{-3} \ M$ we can also disregard reaction (7), we see that only reactions (5), (6), and (8)–(10) should be taken into account.

This simple reaction scheme is consistent with measured radiation-chemical yields of stable radiolytic products. In this example we shall show in more detail how we derive kinetic expressions which relate the radiation-chemical yields of the primary products of water radiolysis to be determined to measured radiation-chemical yields of stable products of the radiolysis of the system studied. Here we have a case of steady-state kinetics where it is assumed that each free radical produced by water decomposition must in a way also disappear, that is, that a steady concentration of the radical is established. It depends on the dose rate, as well as on the concentrations of all scavengers and the rates of their reactions with the radical observed. Instead of writing a number of differential equations that are in common use in chemical kinetics, we shall make use of a simpler procedure [1]: It is assumed that the total number of radicals of a given species produced per absorbed 100 eV of radiation energy by reactions in the proposed scheme must disappear by some other reactions of the scheme. We denote the number of times a reaction occurs per 100 eV by using the number of the reaction put in parentheses. Then, taking into account primary yields, we have

for H $\qquad G_H = (5)$
for e_{aq}^- $\qquad G_{e_{aq}^-} = (6)$
for OH $\qquad G_{OH} = (8)$
for HO_2 or O_2^- $\qquad (5) + (6) + (9) = 2(10)$
for HCOO or COO^- $\qquad (8) = (9)$

In this case, the measured yield of formation of carbon dioxide is

$$G(CO_2) = (9) = (8) = G_{OH} \tag{11}$$

The yield of molecular hydrogen is

$$G(H_2) = G_{H_2} \tag{12}$$

while that of hydrogen peroxide is

$$G(H_2O_2) = G_{H_2O_2} + (10) = G_{H_2O_2} + 0.5[(5) + (6) + (9)]$$
$$= G_{H_2O_2} + 0.5[(5) + (6) + (8)]$$
$$G(H_2O_2) = G_{H_2O_2} + 0.5(G_H + G_{e_{aq}^-} + G_{OH}) \tag{13}$$

The equation of material balance,

$$G_{OH} + 2G_{H_2O_2} = G_H + G_{e_{aq}^-} + 2G_{H_2} \qquad [3]$$

used with Eqs. (12) and (13), gives, upon some rearrangements,

$$G(H_2O_2) - G(H_2) = G_H + G_{e_{aq}^-} \tag{14}$$

Hence, Eqs. (11) and (14) can be used to calculate free-radical yields. The yield of primary molecular hydrogen is given by Eq. (12). The yield of primary hydrogen peroxide is calculated from Eq. (3).

III. THE G_R AND G_M VALUES

B. Measured Yields of the Stable Products of the Radiolysis of Sodium Formate in the Presence of Oxygen—Calculation of Corrections for Increased Reactivity and of G_R and G_M

Figure 5.2 shows how an increase in the concentration of formate affects measured yields [14,15]. These experimental yields were not used directly in the above expressions [Eqs. (11), (12), and (14)], but were corrected beforehand for scavenging in the spur and, when necessary, also for the contribution of reaction (7) to the H_2 and CO_2 formation. For calculating the corrections for scavenging within the spur, the authors [14] chose, with some minor modifications, the method given by Flanders and Fricke [37], as used by Fielden and Hart [54,55]. It was assumed that

$$G(CO_2)_{corr} = G(CO_2)_{meas} \times 0.551/I_S$$

where I_S is the fraction of radicals combining with the solute S. The values of $0.551/I_S$ were taken from Table 1 of Flanders and Fricke [37] for $E = 2.5$ and the corresponding values of B, a dimensionless parameter derived from the expression

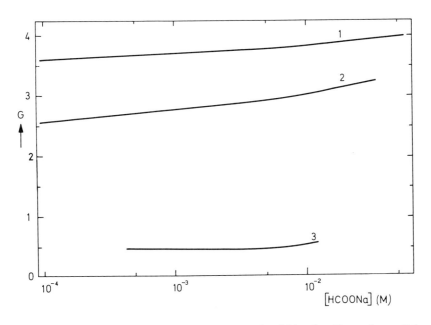

Fig. 5.2. Effect of formate ion concentration on the yields of stable products of the γ-radiolysis of neutral aqueous solutions of HCOONa + O_2. Curve 1 represents the H_2O_2 yield, curve 2 the CO_2 yield, and curve 3 the H_2 yield. (After Draganić et al. [14] and Hart [15].)

$$B = \frac{b^2}{4D} k_{\text{OH}+\text{S}} \times [\text{S}]$$

Here, $b = 5.8$ Å is the radius of the spur for OH radicals (considered to be 2.7 times smaller than that for e_{aq}^- [33,56]), $D = 2 \times 10^{-5}$ cm^2 sec^{-1} is the diffusion coefficient for the OH radical [56], $k_{\text{OH}+\text{S}}$ is the rate constant for reaction (8), and [S] is the formate ion concentration in moles per liter. Substituting these values into the above expression, we obtain

$$B = 1.05 \times 10^{-1} \times [\text{HCOO}^-]$$

where the concentration is expressed in moles per liter.

The $G(\text{CO}_2)$ values measured at $[\text{HCOO}^-] \geqslant 1 \times 10^{-3}$ M are also corrected for reaction (7). The oxygen concentration (1×10^{-3} M) is such that reactions (5) and (6) also proceed to some extent in the spur. As these affect the yields of H_2O_2, corrections were made before inserting the measured values into Eq. (14). The corrections were calculated in the manner described above. The parameter B was calculated as

$$B = 2.44 \times [O_2]$$

by taking the data for the hydrated electron [54]: $b = 15.8$ Å and $D = 4.8 \times 10^{-5}$ cm^2 sec^{-1}. The oxygen concentration is expressed in moles per liter.

Table 5.1 presents experimental values for the stable product yields and

TABLE 5.1

Yields Measured in γ-Radiolysis of HCOONa + O_2 and Corrected for Increased Reactivity[a,b]

HCOONa, M	$G(CO_2)$			$G(H_2O_2)$			$G(H_2)$	
	A	B	C	A	B	C	A	C
1×10^{-4}	2.48[c]			3.69[c]				
5×10^{-4}	2.56	2.53	2.56	3.68	3.59	3.60	0.43	0.44
1×10^{-3}							0.43	0.43[d]
5×10^{-3}	2.98	2.85[d]	2.82[d]	3.78	3.69	3.64	0.50	0.48[d]
1×10^{-2}	2.90[e]	2.72[d]	2.74[d]	3.60[e]	3.51	3.46	0.46[e]	0.45[d]
2.5×10^{-2}	3.22	2.82[d]	2.72[d]					
5×10^{-2}				3.87	3.78	3.56		

[a] After Draganić et al. [14].
[b] Legend: A, measured yield; B, measured yield corrected according to Flanders and Fricke [37]; and C, measured yield corrected according to Kuppermann [56].
[c] No correction, low hydroxyl radical reactivitity.
[d] Corrected also for the contribution of reaction (7).
[e] Value taken from Hart [15].

the values obtained after introducing the corrections described above. It also gives the measured values corrected according to Kuppermann's diffusion-kinetic theoretical curves [56] (see Chapter Six).

The corrections derived from Kuppermann's curves on diagrams are less precise than those calculated by the Flanders and Fricke method [37]. This may be especially true in the relatively low reactivity region studied in that work [14], where the effect of scavenging within the spur is not very appreciable. However, Kuppermann's curves offer greater possibilities of estimating the various corrections as derived from a more elaborate model of water radiolysis (seven primary species in a mechanism involving twenty reactions). The need for the different corrections can be seen from the following considerations: The increase in formate ion concentration leads not only to an apparent increase of G_{OH} but, by preventing the hydroxyl radical recombination, also to an apparent decrease in $G_{H_2O_2}$. The efficient scavenging of primary reducing species by oxygen leads not only to an apparent increase of G_{red}, but also to a corresponding decrease of G_{H_2}. Furthermore, higher reactivities toward hydrated electrons (1.9×10^7 sec^{-1} in the presence of 1×10^{-3} M oxygen) and hydroxyl radicals (in the above experiments up to 1×10^8 sec^{-1}) prevent the water-forming reaction between OH and e_{aq}^- (or H) in the spur and leave in excess a certain number of primary oxidizing and reducing species, respectively.

As can be seen, the measured yields (column A) corrected by the two methods (columns B and C) are in fairly good agreement, although in some cases the corrections differ quite considerably from each other. This is not surprising, as in both cases the corrections are relatively small compared to the measured yields to which they are applied. The mean values, obtained by correcting the measured yields in the two ways mentioned above, were used in the primary yield calculations, which give

$$G_{-H_2O} = 4.09 \qquad G_H + G_{e_{aq}^-} = 3.18 \qquad G_{OH} = 2.72$$
$$G_{H_2} = 0.45 \qquad G_{H_2O_2} = 0.68$$

C. Measurements in Some Other Systems

It is certain that improvement in the diffusion radical model will lead to some changes in the parameters used in calculating the corrections, and hence also to some changes in the primary yields derived in the case treated in the preceding section. Therefore, it is interesting to consider some other measurements where it was not necessary to use such corrections or where these were small.

Using a sensitive spectrophotometer (the Cary Model 16), Bielski and Allen [57] measured peroxide formation in air-saturated 5×10^{-4} and 1×10^{-3} M solutions of sodium formate, at absorbed doses between 30 and

150 rad only. The hydrogen peroxide yield under these conditions is 3.74, and it does not differ appreciably from the value given in the preceding section. An advantage of these measurements, however, is that they were done in very dilute solutions, where no corrections for reactivity should be introduced, and at extremely low absorbed doses, where secondary reactions are negligible. The authors also obtained the same value for $G(H_2O_2)$ in air-saturated 1.7×10^{-3} M solutions of ethanol, at absorbed doses lower than 300 rad. Since in this case the reaction scheme is also the same as that considered for sodium formate in Section I.A, these data also allow one to calculate, by means of Eq. (14), the total yield of primary reducing species: $G_H + G_{e_{aq}^-} = G(H_2O_2) - G(H_2) = 3.34$. Here, it is assumed that $G(H_2) = G_{H_2} = 0.40$. In the same work, the yield of oxidation of ferrous ions was also measured in an air-saturated solution at pH 5.8. The reaction scheme is made up of the reactions given by Eqs. (5) and (6), as well as of the following:

$$Fe^{2+} + OH \longrightarrow Fe^{3+} + OH^- \tag{15}$$

$$Fe^{2+} + H^+ + HO_2 \text{ (or } O_2^-) \longrightarrow Fe^{3+} + H_2O_2 \tag{16}$$

$$Fe^{2+} + H_2O_2 \longrightarrow Fe^{3+} + OH^- + OH \tag{17}$$

In this case, steady-state kinetics gives

$$G_H + G_{e_{aq}^-} = \tfrac{1}{4}[G(Fe^{3+}) - 2G_{H_2}] \tag{18}$$

Since it was found that $G(Fe^{3+}) = 13.95$, we calculate that

$$G_H + G_{e_{aq}^-} = 3.29$$

These authors also used the experimental value for $G(H_2O_2)$ in ethanol and formate to calculate the yield of hydroxyl radicals. Here, they made use of a previous finding [58] that the value of $G(H_2O_2)$ in air-saturated water containing only a small quantity of bromide is 1.00 and is given by

$$G(H_2O_2) = G_{H_2O_2} + 0.5(G_H + G_{e_{aq}^-} - G_{OH}) \tag{19}$$

Subtracting Eq. (19) from (13), they obtained $G_{OH} = 2.74$.

Haissinsky was among the first [59] to point out that the radical yields given above are more likely than the substantially lower values which were usually measured in a neutral medium (see Section IV.B). Studying the radiolysis of neutral solutions of phosphite in the presence of nitrate, he assumed that the nitrite and molecular hydrogen are formed by the following mechanism:

$$HPO_3^{2-} + OH \longrightarrow HPO_3^- + OH^- \quad k_{20} = 1.9 \times 10^9 \tag{20}$$

$$HPO_3^{2-} + H + H_2O \longrightarrow HPO_3^- + H_2 + OH^- \tag{21}$$

$$HPO_3^- + OH \longrightarrow H_2PO_4^- \tag{22}$$

III. THE G_R AND G_M VALUES

$$HPO_3^- + H_2O_2 \longrightarrow H_2PO_4^- + OH \quad (23)$$
$$NO_3^- + e_{aq}^- + H_2O \longrightarrow NO_2 + 2OH^- \quad (24)$$
$$NO_2 + HPO_3^- + H_2O \longrightarrow NO_2^- + H_2PO_4^- \quad (25)$$

From this it follows that

$$G_{e_{aq}^-} = G(NO_2^-) \qquad G(H_2) = G_{H_2} + G_H$$

On the basis of the experimentally determined yields of molecular hydrogen, he concluded that for $G_{H_2} = 0.45$, $G_H = 0.55$. From the nitrite yield, he calculated $G_{e_{aq}^-}$ to be 2.65.

The results of some measurements in the system $CO + O_2$ [60] are worth attention. The reaction scheme is the same as in the case of the system formic acid–oxygen: Oxygen removes the reducing species by reactions (5) and (6); CO reacts with OH radicals to give the carboxyl radical which disappears by reaction (9); and hydrogen peroxide is formed by reaction (10). It was found that $G(H_2O_2) = 3.62$ in solutions in which [CO] was 6.8×10^{-5}, 14.5×10^{-5}, and 32.8×10^{-5} M, and $[O_2] \sim 1 \times 10^{-3}$ M. Assuming that $G(H_2) = G_{H_2} = 0.45$ and correcting the value of peroxide yield for intraspur reactions, the authors determined $G_H + G_{e_{aq}^-}$ to be 3.13. They also measured $G(H_2O_2)$ to be 1.03 in oxygen-swept dilute KBr solution (2.5×10^{-5}) M for which Eq. (19) holds. Subtracting Eq. (19) from Eq. (13) and substituting the measured peroxide yields, they obtained $G_{OH} = 2.59$.

Sawai [61] irradiated deaerated neutral solutions of 2-propanol (2.5×10^{-2} M) containing a scavenger for hydrated electrons. Here, we have

$$(CH_3)_2CHOH + H \longrightarrow (CH_3)_2COH + H_2$$
$$k_{26} = 2 \times 10^7 \ M^{-1} \ \text{sec}^{-1} \quad (26)$$

$$(CH_3)_2CHOH + OH \longrightarrow (CH_3)_2COH + H_2O$$
$$k_{27} = 1.7 \times 10^9 \ M^{-1} \ \text{sec}^{-1} \quad (27)$$

If the irradiated solutions contain hydrogen peroxide as the scavenger for hydrated electrons, but at sufficiently low concentrations (about 1×10^{-3} M) for reaction (26) to be undisturbed, then $G(H_2) = 1.06$ is in fact a measure of $G_H + G_{H_2}$. If the irradiated solution contains KNO_3, which effectively removes e_{aq}^-, then as its concentration increases, the measured hydrogen yield decreases. However, in this case also the value $G(H_2) = 1.06$ is obtained for conditions under which reaction (26) proceeds without hindrance. Under the assumption that $G_{H_2} = 0.45$, from these measurements it follows that $G_H = 0.6$. Here, it should be noted that, measuring the nitrite and acetone produced, the author also calculated $G_{e_{aq}^-}$ and G_{OH} to be 2.77 and 3.0, respectively; these values should be taken with reserve, in view of the complexity of the reaction scheme and the relatively high reactivity of the system toward hydroxyl radicals.

Baxendale and Khan [62] irradiated very dilute (6×10^{-6}–3×10^{-5} M) neutral solutions of p-nitrosodimethylaniline (PNDA) with pulses of only 15 and 40 rad; hence, under very initial conditions where only the reactions of PNDA with primary species are to be taken into account. These measurements were carried out under different conditions. Assuming that in the case of argon-saturated solution all free radicals are effectively removed, we have

$$G(-\text{PNDA}) = G_\text{H} + G_{e_\text{aq}^-} + G_\text{OH}$$

This relation also holds for solutions that are at the same time saturated with N_2O. This last effectively converts hydrated electrons into hydroxyl radicals (Chapter Four), but this should not affect the measured yield of PNDA decomposition. Indeed, we see that there is no essential difference, since in the first case $G(-\text{PNDA})$ is measured to be 6.22, and in the second case it is 6.47. The somewhat higher value in the case of nitrous oxide is accounted for by a higher reactivity toward hydrated electrons and by the contribution of species from the spur. In the presence of an effective scavenger for hydrated electrons such as $Co(NH_3)_6^{3+}$, where the product obtained does not react with PNDA, the following relation should hold

$$G(-\text{PNDA}) = G_\text{H} + G_\text{OH} = 3.39$$

In solutions which were saturated with oxygen, so that both the hydrated electron and the H atom were effectively removed, and under the assumption that the hydroperoxyl radical produced does not react with PNDA, it was found that $G(-\text{PNDA}) = G_\text{OH} = 2.75$. From these data the authors calculated

$$G_\text{OH} = 2.75, \quad G_\text{H} = 0.64, \quad \text{and} \quad G_{e_\text{aq}^-} = 2.83$$

The dose in pulse was calibrated by an oxygen-saturated Fricke dosimeter, but the accuracy of the dosimetry was not indicated.

Pulse radiolysis experiments [54] gave $G_{e_\text{aq}^-} = 2.65$ for pH 7; in this case special attention was paid to dosimetry.

Competition experiments [63] gave $G_\text{H} = 0.48$. Here, use was made of formate, acetate, or acetone as the scavengers for H atoms. In this case, the reaction scheme is in the general form

$$\text{H} + \text{HR} \longrightarrow \text{H}_2 + \text{R}$$
$$e_\text{aq}^- + \text{S} \longrightarrow \text{P}$$

where S is acetone or nitrate. It is assumed that the hydroxyl radicals effectively disappear by the reaction

$$\text{OH} + \text{HR} \longrightarrow \text{H}_2\text{O} + \text{P}'$$

where P′ does not react with H atoms. The reactivities (v, in reciprocal seconds) are adjusted so that $v_{\text{H}+\text{HR}} \gg v_{\text{H}+\text{S}}$ and $v_{e_\text{aq}^- + \text{S}} \gg v_{e_\text{aq}^- + \text{HR}}$, and the measurements reduce to the determination of molecular hydrogen.

In similar experiments [64], the solute S was chosen so that it reacted very efficiently not only with hydrated electrons but also with H atoms, and that no molecular hydrogen was produced. For HR the same holds as in the preceding case. Hydroxyl radicals were removed by both S and HR, but no molecular hydrogen was formed. As the HR, use was made of methanol, ethanol, benzoquinone, isopropanol, and formate, while nitrates, nitrites, Cd^{2+}, Zn^{2+}, Ni^{2+}, OH^-, Cu^{2+}, Ag^+, Hg^{2+}, ferricyanide, chromate, and dichromate were used as the scavenger S. From these experiments it follows that the values of G_H in neutral media, derived from competition diagrams, range from 0.47 to 0.70, depending on the system.

Mahlman [65] tried to avoid conclusions based on competition experiments and made use of an approach which in principle ensures more reliable conclusions. He measured $G(H_2)$ in neutral aqueous solutions of N_2O also containing sodium nitrate and 2-propanol or sodium formate. The molecular hydrogen measured is due to G_{H_2}, the reaction H + S, and "direct" action on S. He found G_H to be 0.57 and 0.61.

Very often for measurement of primary molecular hydrogen in irradiated water of neutral pH it suffices to avoid the reaction $H_2 + OH$ for $G(H_2)$ to be a measure of G_{H_2}. Since this reaction is relatively slow, low concentrations of the scavenger for OH radicals (about 10^{-4} M bromide, iodide, or nitrite) are sufficient. Such measurements [66.67] give $G_{H_2} = 0.45$. The same value was found in other systems [28,31,40]. In these studies, molecular hydrogen was measured at various solute concentrations and the measured yields were plotted against $[S]^{1/3}$, where [S] was the concentration of the solute; G_{H_2} was taken as the value corresponding to the zero solute concentration.

It is more difficult to apply the same approach to the measurement of $G_{H_2O_2}$, because hydrogen peroxide is both chemically and radiation-chemically more reactive than molecular hydrogen—hence it is more difficult to protect. Measurements in dilute deaerated solutions ($\sim 10^{-4}$ M) of ethanol, acrylamide, potassium nitrate, and acetone at pH ~ 6 give $G_{H_2O_2} = 0.67 \pm 0.01$ [35]. It is worth recalling the experiments performed by Backhurst et al. [68]. The authors irradiated water in which all oxygen was ^{18}O ($H_2^{18}O$) and to which the following scavengers were added: ethanol for OH radicals and O_2 (natural isotopic composition) for reducing species. Under these conditions, the peroxide measured was produced by reactions similar to those for oxygenated solutions of formic acid [Eqs. (5), (6), (9), and (10)], and relation (13) $[G(H_2O_2) = G_{H_2O_2} + 0.5(G_H + G_{e_{aq}^-} + G_{OH})]$ holds for it. If the assumption that the $G_{H_2O_2}$ is due to the recombination of hydroxyl radicals from the water (^{18}OH) is valid, then the percentage of ^{18}O in the total oxygen from the hydrogen peroxide shows us the yield of the primary H_2O_2. This is possible because the rest of the hydrogen peroxide is produced by reactions of O_2 with reducing species and with the organic radical, and this oxygen is

TABLE 5.2

Yields of the Primary Radical and Molecular Products of Water γ-Radiolysis at Neutral pH, Derived from Various Systems

$G_{e_{aq}^-} + G_H$	$G_{e_{aq}^-}$	G_H	G_{OH}	G_{H_2}	$G_{H_2O_2}$	Reference
3.18			2.72	0.45	0.68	[14, 15]
3.31			2.74			[57]
	2.65	0.55		0.45		[59]
	2.65					[54]
3.13			2.59		0.72	[60]
	2.83	0.64	2.75			[62]
		0.61				[61]
		0.48				[63]
		0.61; 0.57				[65]
				0.45		[28, 31, 40, 66, 67]
					0.67	[35]
					0.7	[30, 58]

known to have the natural isotopic composition. The value of $G_{H_2O_2}$ calculated by the authors (0.73) is somewhat higher than the true yield. It was computed on the basis of the experimental value $G(H_2O_2) = 3.2$, which is at present known [57] to be substantially lower than the true initial yield in this system.

Table 5.2 summarizes the yields of primary species in the γ-radiolysis of neutral water reported in the studies which were considered in this section.

IV. Effect of pH

A. Some Measurements over a Wide pH Range

In Chapter Three we have seen that in an acid medium, e_{aq}^- may be effectively converted into the H atom by the reaction

$$e_{aq}^- + H_3O^+ \longrightarrow H + H_2O \qquad (28)$$

The conversion of the hydroxyl radical into the O^- ion radical,

$$OH + OH^- \longrightarrow O^- + H_2O \qquad (29)$$

was also considered in some detail in Chapter Four. Since the species produced by the conversion have different diffusion coefficients and, to a certain extent, different recombination rate constants, it is evident that the relative

IV. EFFECT OF pH

abundance of H compared to e_{aq}^-, and of OH compared to O^-, may affect the G_R and G_M values at extreme pH's. As we shall see in Section IV.B, many earlier studies indicated that this pH effect should be considerable. It seems very likely that this is not the case and that water decomposition yields do not essentially depend on whether or not reactions (28) and (29) are involved in the radiolysis mechanism.

The oxidation of ferrous ions in the presence of oxygen was the subject of numerous studies in connection with the effect of acidity on the primary yields of water radiolysis. From the study [57] which we have considered in some detail in Section III.C it is seen that $G(Fe^{3+})$ increases with increasing acidity, from 13.95 to 15.52 when the pH is lowered from 5.8 to 1.3. Also in this case, the radiolysis mechanism involves reactions (5), (6), and (15)–(17). Since the measurements were done in dilute solutions and at very low absorbed doses, no corrections to measured values were introduced. The yield of primary reducing species is calculated from Eq. (18) under the assumption that $G_{H_2} = 0.4$. It was found that $G_H + G_{e_{aq}^-}$ increases from 3.3 at pH 5.8–3.4 to 3.65 at pH ≤ 1.3.

Boyle [69] studied the effect of acidity on the oxidation of ferrous ions over the range of very high sulfuric acid concentrations from 0.4 to 18 M. Here the reaction scheme is much more complex than in the case considered above. It seems that further studies are needed to better understand the effect of strong acidities on the primary yields.

Studying the pH effect in alkaline media, Czapski and Peled [70] observed alkaline N_2O saturated solutions of 2-D-2-propanol and of ferricyanide–ferrocyanide mixtures. The oxygen formed in the N_2O–ferricyanide–ferrocyanide system at pH > 11 was taken as a measure of $G_{H_2O_2}$:

$$2Fe(CN)_6^{3-} + H_2O_2 \longrightarrow 2Fe(CN)_6^{4-} + 2H^+ + O_2 \qquad (30)$$

From these results it follows that at pH between 11 and 13, $G_{H_2O_2} = 0.65 \pm 0.05$, and thereafter it begins to decrease. At pH 13.5 the decrease amounts to about 20%, whereas at pH 14 the measured value is reduced by 50%. The $G(N_2)$ variations with pH and ferricyanide concentration were used to calculate $G_{e_{aq}^-}$ and G_H, assuming different values for the rate constants of reactions observed. These calculations give $G_H + G_{e_{aq}^-} = 3.3$ at pH 11–14. Alkaline solutions of N_2O saturated 2-D-2-propanol were used for the determination of G_H by $G(HD)$ and $G(H_2)$ measurements; these products were formed in the following reactions:

$$H_3C-CDOH-CH_3 + H \longrightarrow HD + H_3C-\dot{C}OH-CH_3 \qquad (31)$$

$$H_3C-CDOH-CH_3 + H \longrightarrow H_2 + H_3C-CDOH-\dot{C}H_2 \qquad (32)$$

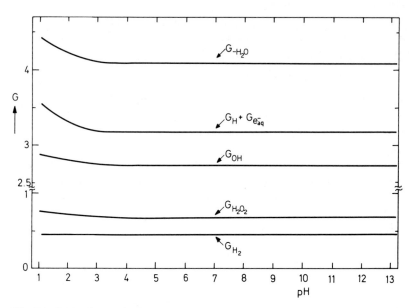

Fig. 5.3. Dependence of primary radical and molecular yields of water γ-radiolysis on pH, derived from measurements on formic acid–oxygen solutions. (After Draganić et al. [*14*].)

At pH 12.9, G_H was found to be 0.6 ± 0.1. Thus, the results obtained show that there is no pH dependence of G_H and $G_{e_{aq}^-}$ when the pH passes from neutral to alkaline.

The radiolysis of $HCOOH + O_2$ was used to calculate all primary yields of water radiolysis over the pH range 1.3–13 [*14,15*]. We have considered in detail (Section III) the case of neutral medium. For acid and alkaline media, the effect of concentration on measured stable radiolytic products has also been studied and corresponding corrections for increased reactivity have been introduced [*14*]. The reaction scheme for acid medium is the same as that for neutral medium. The observed increase in $G(CO_2)$ and $G(H_2O_2)$ points out that reaction (28) leads to a small increase in primary yields, as is seen from Fig. 5.3. The fact that the yields of CO_2 and H_2O_2 do not change with the transition from neutral to alkaline medium led the authors to the assumption that the conversion of the hydroxyl radical by reaction (29) is followed by the reactions

$$HCOO^- + O^- \longrightarrow OH^- + COO^- \qquad (33)$$

$$O_2 + O^- \longrightarrow O_3^- \qquad (34)$$

Since k_{34} is known ($2.6 \times 10^9 \ M^{-1} \ sec^{-1}$), the authors tried to determine k_{33}

IV. EFFECT OF pH

by varying the ratio of oxygen-to-formate concentration. The absence of competition at $[O_2]/[HCOO^-]$ between 0.2 and 8 pointed out the possibility that the following reaction takes place:

$$HCOO^- + O_3^- \longrightarrow COO^- + HO_3^- \tag{35}$$

Since $HO_3^- \to OH^- + O_2$ [71], the sequence of reactions (34) and (35) gives the same result as reaction (33). Similar conclusion would be drawn by taking into account the finding [72] that decay of ozonide ion occurs through thermal dissociation and that O^- is the reactive species. Analysis shows that no other reactions of O^- species may occur under these conditions, but the authors nevertheless think that the primary yields calculated for pH 12–13 should be regarded with some reserve as the rate constant and the products of reactions (33) and (35) are not completely established.

It should be noted that $G_{H_2O_2}$ values were calculated in the above case by means of the equation of material balance [Fq. (3)]. Only a few direct determinations, that is, such determinations where $G(H_2O_2) = G_{H_2O_2}$, have been carried out so far. Such measurements [35] at pH 1.3 in dilute solutions (10^{-5}–10^{-4} M) of ethanol, acrylamide, and 1-propanol gave $G_{H_2O_2} = 0.76 \pm 0.01$. In the same work, measurements in acrylamide solution at pH 13 gave a value of 0.56 ± 0.01, but the authors stress that further studies are needed for a definitive conclusion to be drawn. These studies should be carried out in a system which also makes possible a simultaneous measurement of G_{OH}. As we have pointed out, reaction (29) probably does not induce any essential change, but for the present leads to an uncertainty; the O^- ion radical produced by the conversion recombines to give peroxide by a reaction whose rate constant is substantially smaller than the corresponding value for the recombination of OH radicals (Chapter Four). Furthermore, the reaction of O^- with hydrogen peroxide is known to be fast, which also makes such a consideration difficult.

It is most often taken that $G_{H_2} = 0.45$ over a wide range of pH, from moderately acid to moderately alkaline ones. It is still not established whether or not these values remain unchanged at the ends of the pH scale, for pH < 2 and pH > 13. From the data reported by Cheek et al. [73], concerning both neutral solutions and solutions in which sulfuric acid concentration increased up to 10 M, it follows that the yield of the primary H_2 decreases with increasing acidity (down to about 0.4 at pH 0). Molecular hydrogen was measured [32] in KNO_2 solutions at different concentrations. The values for G_{H_2} were calculated from the intercept of the ordinate of a diagram in which $G(H_2)$ is plotted against $[KNO_2]^{1/3}$. The results of measurements at pH 13, 13.5, and 13.7 do not essentially differ from those in a neutral medium (G values of 0.43 and 0.44). However, the value obtained for pH 14 is substantially lower: $G = 0.37$.

B. *Earlier Studies in Connection with the Dependence of G_R and G_M on pH*

From what we have seen in the preceding section it follows that at pH 3–13 no change in G_R and G_M values occurs. An increase in acidity (pH < 3) undoubtedly leads to a slight increase in the G_R value, as well as to a small increase in $G_{H_2O_2}$, while G_{H_2} under these conditions remains constant or slightly decreases. An increase in alkalinity (pH > 13) most likely does not lead to any appreciable change in G_R, but this, as well as corresponding changes in G_M, should be more fully investigated; in particular, attention should be drawn to the drops of G_M observed in some cases. All in all, the primary yields do not depend very much on pH. This picture of the situation may perplex the reader who tried to arrive at his own conclusion on the basis of previously published studies (more than 100 in the period 1952–1967), in which the idea dominated that, in neutral water, G_R values are appreciably lower from those given in Table 5.2 (even by as much as 20%), and that they increase at extreme pH's. From the analysis of published works given by Haissinsky [2] it can be seen that answers may considerably differ from each other, depending on the system from which they are derived. Thus, for neutral pH, G_{OH} values vary between 2.1 and 2.7, and those of $G_H + G_{e^-_{aq}}$ vary between 2.7 and 3.1; similar discrepancies are also encountered for alkaline media, where G_R values were often larger by as much as 30% than those for neutral medium. Hayon [3] tried to explain the disagreement between the results obtained with different systems by the fact that the effect of reactivity was disregarded in some cases. Moreover, he ascribed the increase in G_R at the ends of the pH range to the increased reactivity due to the occurrence of reactions (28) and (29), respectively. This approach was disputed on theoretical as well as experimental grounds by Czapski [4]. He concluded that in many cases where the G values of radicals seemed to depend on pH or on scavenger concentration, the effect was in reality an artifact resulting from disregard of the effect of back reactions. His general conclusion, as well as that of Haissinsky [2], is the same as that given above: There is no strong dependence of the primary yields on pH, although the situation at extreme pH's is not yet quite clear. In this connection it should be noted that the effect of extremely high acidities or alkalinities (multimolar solutions of acids and NaOH, respectively) represents a special problem which still requires a particular study.

V. Primary Yields in D_2O

It is surprising how little attention has been paid to the determination of the primary yields in the radiolysis of D_2O: G_{D_2}, $G_{D_2O_2}$, G_{OD}, G_D, and $G_{e^-_{aq}}$. Discrepancies between a few measurements which have been done so far are considerable, and they reflect all the weak points which we have mentioned before in connection with the determination of G_R and G_M in H_2O (Section I.B). Therefore, we shall dwell not on the absolute values of the yields but on their ratios for light and heavy water, derived from measurements with the same system and under the same experimental conditions. It should be noted that we are interested only in the radiolysis of heavy water and not in phenomena occurring in D_2O–H_2O mixtures. For the present, these latter are of interest mainly for those who are studying the isotope effect, for which the papers of Lifshitz and Stein [74] and Anbar and Meyerstein [75] give quite a complete review.

Let us see what the difference is in the physical properties of D_2O and H_2O and what we might expect on the basis of our present knowledge of the interaction of radiation with matter (Chapter Two). At room temperature, for D_2O the mass density is higher by 11 % than that of H_2O, but the electron density is only 0.3 % higher and there is no essential difference in the amount of absorbed radiation energy in a given volume of the two media. Furthermore, it also seems that there is no appreciable difference in the number of excited and ionized molecules under the same irradiation conditions; on the basis of available data on excitation potentials it is generally accepted [74,76,77] that these are practically the same for D_2O and H_2O. However, it should be noted that a small difference might nevertheless be expected. Certain theories of the structure of liquid water [78,79] indicate an increase in the hydrogen bond strength of D_2O compared to that of H_2O. Also, mass spectrographic measurements showed that molecules containing a heavy isotope are more difficult to dissociate than corresponding molecules containing a light isotope. Platzman [80] suggests that such molecules are more susceptible to preionization. In the case of water, this might mean that the radiation yield of the hydrogen produced by dissociation of excited water molecules is lower in D_2O than in H_2O, that is, that $G_D < G_H$. Such an effect should be quite small, only a few percent [74].

The relaxation time in D_2O is longer than that in H_2O. This might lead to a difference in primary yields [77], but not necessarily [76]. As a matter of fact, the prolonged relaxation time would allow the ejected electron to reach a larger distance from the D_2O^+ than from the H_2O^+ before becoming hydrated. The larger the distance reached, the smaller is the number of initial species which recombine, hence, the higher are their yields and the lower are the yields of corresponding molecular products. Decomposition of heavy water might be somewhat higher if the increase in G_R is not compensated for by a corresponding decrease in G_M.

Hardwick [76] determined stable radiolytic products in the system $HCOOH + O_2$ in light and heavy water. The mechanism of this process and the calculation of primary yields for measured radiolytic products were considered in Section III. Table 5.3 shows the ratios $(G_R)_{H_2O}/(G_R)_{D_2O}$ and $(G_M)_{H_2O}/(G_M)_{D_2O}$ as calculated from these experimental data, together with the ratios as inferred from other works. Armstrong et al. [77] irradiated solutions of acrylamide in H_2O or D_2O, in which the acidity was adjusted by perchloric acid ($\sim 4 \times 10^{-2}$ M). The hydrogen peroxide and molecular hydrogen produced were determined as a measure of corresponding primary yields. In a number of experiments, the solutions contained ferric ions at different concentrations. In these cases the reduction yield was a measure of free radicals due to the reaction of the polyacrylamide radical (m_j):

$$(m_j) + Fe^{3+} \longrightarrow P_j + Fe^{2+} \tag{36}$$

where P_j is the molecule of the polymer with j unit monomers. Since polyacrylamide radicals (m_j) are inert with respect to molecular products, we have

$$G_{H_2} = G^0(H_2) \tag{12}$$

$$G_{H_2O_2} = G^0(H_2O_2) \tag{37}$$

where G^0 is the yield measured in the absence of ferric ions. Furthermore, by the usual kinetic treatment we find

$$G_H + G_{e^-_{aq}} = 0.5G(Fe^{2+}) + G(H_2O_2) - G(H_2) \tag{38}$$

$$G_{OH} = 0.5G(Fe^{2+}) - G(H_2O_2) + G(H_2) \tag{39}$$

$$G_{-H_2O} = G(H_2) + G(H_2O_2) + 0.5G(Fe^{2+}) \tag{40}$$

Taking into account the dependence of measured stable radiolytic products in this system on scavenger concentration, we calculate by means of the above equations the ratios given in Table 5.3. A number of measurements were carried out in specially chosen systems at different pD (or pH) values between 0 and 14 [81]. The molecular hydrogen was measured in degassed solutions of KBr and of KNO_3. According to the reaction scheme, the measured yield of hydrogen is equal to that of the primary molecular hydrogen. The data on $G_{D_2O_2}$ were derived from measurements in oxygenated solutions of KBr or ceric sulfate, as well as in degassed solutions of a ferrocyanide–ferricyanide system. The G_D value was calculated from measured hydrogen yield in the system ethanol–KNO_3, giving $G_H/G_D = 1.15$. The total yield of reducing radicals was calculated from measurements in degassed solutions of ethanol as well as in solutions of KNO_3 + HCOONa. Oxygenated solutions of ferricyanide + HCOONa were used to determine G_{OD} values. It should be noted that all the G values given in reference [81] for heavy water solutions are lower by 11% [3] and were corrected before they were used in calculating the data given in Table 5.3. Draganić et al. [82] derived the data from various combinations of the following scavengers: Cu^{2+}, Br^-, oxalic acid, ethanol, and oxygen, over the pD and pH range 1.3–13. It is interesting that the measured value for G_H/G_D (=1.1) supports what was said above about dissociation

V. PRIMARY YIELDS IN D_2O

of heavy and light water, but the authors point out that the possible experimental error in this case may be larger than the effect. The ratios of the yields of primary species which the authors calculated from their measurements are given in Table 5.3. Fielden and Hart [55] studied the pulse radiolysis of heavy water solutions over the pD range 7–14 and compared the values derived here with those obtained in their experiments in light water [54]. It is interesting that here $G_H/G_D > 1$ and $(G_R)_{H_2O}/(G_R)_{D_2O} < 1$, as was found in the preceding work. In view of the fact that there are still some doubts as to the origin of the hydrogen atoms, the data on G_H/G_D may be a useful argument that they are also produced by dissociation of H_2O^*. Coatsworth et al. [83a] measured the yields of oxidation of ferrous ion in acid solutions (0.8 N H_2SO_4) irradiated in the presence and absence of oxygen. In the first case, molecular hydrogen was also measured. According to the oxidation mechanism, the yields of the products measured are

$$G^0(Fe^{3+}) = G_D + G_{e_{aq}^-} + G_{OD} + 2G_{D_2O_2} \tag{41}$$

$$G(Fe^{3+}) = 3(G_D + G_{e_{aq}^-}) + G_{OD} + 2G_{D_2O_2} \tag{42}$$

$$G(D_2) = G_{D_2} \tag{12a}$$

where G^0 denotes the yield measured in the absence of oxygen. The data calculated from experimental values and these equations are given in Table 5.3. It should be noted that $G(Fe^{3+})$ was determined here for the Fricke dosimeter in heavy water to be 16.2, whereas for light water the same authors give 15.07. For the yield of ferric ions in a heavy water Fricke dosimeter we also find the following values in the literature (the corresponding value accepted for the yield in light water is given in parentheses): 16.95 (15.5) [76], 16.72 (15.6) [67], and 16.5 (15.5) [82]. These measurements also point out the increase of the yield of free radicals in heavy water. Mahlman and Boyle [67] measured the hydrogen produced in degassed acidic (0.8 N H_2SO_4) solutions of KBr (10^{-4} M) where, according to the reaction scheme, $G(D_2) = G_{D_2}$. Hydrogen peroxide was measured in corresponding aerated aqueous solutions and calculated from the equation

$$G(D_2O_2) = G_{D_2O_2} + 0.5(G_D + G_{e_{aq}^-} - G_{OD}) \tag{19a}$$

The yield of ferric ions in acidic solutions of ferrous sulfate (0.8 N H_2SO_4), which relates primary radical yields to $G_{H_2O_2}$ by the expression already given in the preceding case, was also measured. The experimental yields and the above relations allow us to calculate all primary yields by means of the equation of material balance. The yield of hydrated electrons in H_2O and D_2O has been derived from measurements in corresponding solutions of sulfur hexafluoride [84].

TABLE 5.3

RATIOS OF THE RADIATION-CHEMICAL YIELDS OF PRIMARY PRODUCTS OF THE RADIOLYSIS OF H_2O AND D_2O, $(G_R)_{H_2O}/(G_R)_{D_2O}$ AND $(G_M)_{H_2O}/(G_M)_{D_2O}$, DERIVED FROM VARIOUS SYSTEMS

Total reducing radicals	Hydrated electron	Atomic hydrogen	Hydroxyl radical	Molecular hydrogen	Hydrogen peroxide	Water decomposition	Reference
0.89			0.95	1.46	1.03	0.97	[76]
0.93			0.89	1.39	1.23	0.96	[77]
0.91		1.15	0.96	1.1	1.13	0.99	[81]
0.96	0.92	1.1	0.93	1.23	1.07	1.00	[82]
0.92	0.88	1.25					[55]
0.90			0.95	1.21	1.04	0.97	[83a]
			0.92				[83b]
0.92			0.94	1.20	1.00	0.97	[67]
	0.90						[84]
0.95	0.89	1.28	0.96	1.30	1.03	1.00	[85]

As can be seen in Table 5.3, there is no essential difference between G_{-H_2O} and G_{-D_2O}; practically, for 100 eV absorbed, the same number of light and heavy water molecules are decomposed. Furthermore, it seems that, regardless of pH (or pD), $(G_R)_{H_2O} < (G_R)_{D_2O}$ and $(G_M)_{H_2O} > (G_M)_{D_2O}$, as well as $G_H > G_D$. However, these results should be taken with some reserve. It is known that the data used in deriving the primary yields were not always the true initial yields. Moreover, the reaction schemes used for the calculations often seem to be somewhat more complex than it was supposed in the works cited. These weak points might be the same for H_2O and D_2O solutions, so that they may have no appreciable effect on the accuracy of the conclusions concerning the ratios of the yields. This is why we have insisted on ratios rather than on absolute values. Results [85] obtained with heavy water solutions of formic acid in the presence of oxygen or nitrate are, however, worth noting. In these experiments particular care was taken that the yields of stable products (CO_2, D_2, HD, and D_2O_2) are the true initial yields. Measured values were corrected for scavenging in the spur and, when necessary, for other contributions. As was seen in Section III.A, the reaction scheme also seems to be well established. The absolute values for primary yields from the above system were obtained for pD 1.3–13:

acid medium (pD = 1.3)

$$G_D + G_{e_{aq}^-} = 3.67, \quad G_{OD} = 3.00, \quad G_{D_2} = 0.32,$$

$$G_{D_2O_2} = 0.66, \quad G_{-D_2O} = 4.32$$

V. PRIMARY YIELDS IN D_2O

neutral and alkaline media

$$G_D = 0.43, \quad G_D + G_{e_{aq}^-} = 3.39, \quad G_{e_{aq}^-} = 2.96,$$

$$G_{OD} = 2.84, \quad G_{D_2} = 0.36, \quad G_{D_2O_2} = 0.66, \quad G_{-D_2O} = 4.12$$

Table 5.3 also shows the ratios of the yields in H_2O and D_2O derived from the above data and the corresponding data for H_2O, given in Sections III and IV.

Only scarce information is available on the rate constants of the reactions of primary species with each other or with solutes in D_2O solutions. Hart and Fielden [86] studied the reactions of the hydrated electron in heavy water with e_{aq}^-, D, OD, and D_2O. The values obtained show that the rate constants for reactions of the hydrated electron with D and OD are almost the same as those of e_{aq}^- with H and OH. Only the rate constant of the reaction of the hydrated electron with D_2O is an order of magnitude lower than that of the corresponding reaction in light water. Draganić et al. [82] reported an appreciable isotope effect in a number of relative reaction rate constants, as can be seen in Table 5.4. Here the a value shows how many times the competition reaction studied is faster in light than in heavy water ($a = k_{H_2O}/k_{D_2O}$). The difference includes not only the isotope effect in the diffusion rates of the reacting species but also the isotope effect in the reaction mechanism itself. Thus, for example, the pronounced isotope effect ($a = 2.7$) in the competition of H^+ (or D^+) and Cu^{2+} for hydrated electrons cannot be accounted for only by the differences in mobility of the reacting species in these two media. True, these differences do exist, but they cannot cause an effect greater than 20%. Other moments, such as the difference in dissociation of copper salt in H_2O and D_2O solutions, cannot account for such an appreciable difference in the rate constants. It seems that it is the deuteron transfer in the reaction

$$e_{aq}^- + D_3O^+ = D + D_2O \qquad [28a]$$

TABLE 5.4

Isotopic Effects in the Relative Rate Constants[a]

Free radical	Competing scavengers	pH (or pD)	Calculated relative rate constant ratios $a = k_{H_2O}/k_{D_2O}$
H (or D)	C_2H_5OH and Cu^{2+}	1.3	1.4
e_{aq}^-	H^+ (or D^+) and Cu^{2+}	3–5	2.7
e_{aq}^-	H^+ (or D^+) and SF_6[b]		2.1
OH (or OD)	KBr and C_2H_5OH	1.3	1.3
	KBr and C_2H_5OH	~6	1.5
O^-	$C_2O_4^{2-}$ and O_2	13	1.3

[a] After Draganić et al. [82].
[b] Taken from Asmus and Fendler [84].

that is considerably slower than the corresponding process of protonization in light water. It may also be interesting to note that this effect is close to the kinetic isotope effect in the processes of protonization, which points out that the deuteron transfer from the D_3O^+ ion to the solvated electron is slower than the corresponding proton transfer in light water. The same conclusion has been drawn [84] for observed $a = 2.1$ with the following competing scavengers for hydrated electrons: SF_6 and H_3O^+ or D_3O^+. This is also supported by the observation [86] that the rate constant of the reaction of the hydrated electron with D_2O is lower by as much as an order of magnitude than the corresponding value for light water.

VI. Effect of Linear Energy Transfer

We have seen in Chapter Two that the Bethe equation allows us to calculate the energy loss of charged particles (the stopping power). In practice, the linear energy transfer (LET) is used instead. In a somewhat different way it describes the same phenomenon as the mass stopping power [87]. The advantage is that it can be calculated more simply. To obtain the mean LET, the energy of the particle is divided by the total path length in the medium observed. For example, α particles of an energy of 5.3 MeV have a range in water of 38.9 μ and their mean LET in water is

$$\frac{5.3 \times 10^3 \text{ keV}}{38.9 \ \mu} = 136 \text{ keV } \mu^{-1} = 13.6 \text{ eV Å}^{-1}$$

This calculation, as well as that of stopping power, disregards the complexity of the process. First of all, from what we have said about the interaction of radiation with matter it is evident that for a given medium and a given particle the LET drastically changes in the course of the particle's motion; it increases with decreasing velocity of the particle, so that at the end of its track it is many times larger than at the beginning. Furthermore, a large part of energy is taken away by δ rays and the energy loss is not localized along the track of the primary, incident ray, as the calculation implies. Photons transfer their energy to the medium through secondary electrons of different energies; hence we also have different LET values. Some authors have tried to overcome the difficulties by taking these changes into account in more complex calculations. A consequence is that, for the same radiation and medium, quite different LET values may be encountered in the literature. This should be taken into consideration when the results on the effect of LET obtained by different authors are compared.

A more refined approach was made in a quantitative check of the diffusion-kinetic model of water radiolysis, where the LET of radiation was introduced as the energy-averaged spur density along the track [88].

VI. EFFECT OF LINEAR ENERGY TRANSFER

It is easy to predict qualitatively the effect of LET on the yield of a radiation-chemical reaction, if the water radiolysis model and the spatial distribution of the events considered in Chapter Two are taken into account. The increase of LET leads to the increase of the density of primary events per unit space and unit time. This should favor reactions of recombinations of primary species. The value of G_M is expected to increase, while G_R should decrease. If the product P comes mainly from the solute reactions with free radicals (H, OH, and e_{aq}^-), then $G(P)$ should decrease. This can be seen in Fig. 5.4, which represents the dependence of $G(Fe^{3+})$ on radiation LET [1]. The yields of ferric ions formed concern the aerated ferrous sulfate (10^{-3} M) solutions in sulfuric acid (0.4 M), irradiated with photons or charged particles of various energies. Measurements in the region of LET of about 1 keV μ^{-1} refer to soft X rays. Most of the other measurements were carried out in solutions subjected to charged particles (protons, deuterons, and helions) of different energies produced by the cyclotron. The highest LET value in this diagram, 220 keV μ^{-1}, refers to the radiation of an internal source—the fragments of the nuclear reaction $^{10}B(n, \alpha)^7Li$. Account was also taken of the results of measurements with the α rays from ^{210}Po, as well as with the fragments from the nuclear reaction $^6Li(n, \alpha)T$, whose LET values were calculated to be 90 keV μ^{-1} and 30 keV μ^{-1}, respectively.

In Section III we have seen that the ferrous ions are oxidized by OH, H_2O_2, and hydroperoxyl radicals. In aerated aqueous solutions, HO_2 are

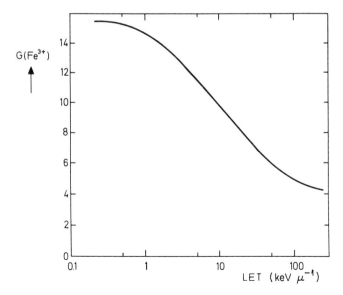

Fig. 5.4. Dependence of ferric ion yields, measured in aerated 1×10^{-3} M ferrous sulfate solutions in 0.4 M sulfuric acid, on LET. (After Allen [1].)

also very efficiently formed in reactions of primary reducing species with oxygen. The overall equation relating the measured ferric ion yields to the yield of primary products of water radiolysis,

$$G(Fe^{3+}) = 3(G_{e^-_{aq}} + G_H) + G_{OH} + 2G_{H_2O_2} \qquad (42a)$$

accounts for the data given in Fig. 5.4. In passing from high-energy photons to charged particles, that is, by increasing LET and favoring the recombination reactions, $G(Fe^{3+})$ decreases as the yields of primary free radicals decrease. It is true that the yield of H_2O_2 is increasing at the same time, but, as is seen in Eq. (42a), its contribution to the total oxidation yield under the above conditions is less important than that of free radicals.

Similar observations have been made in some other systems. However, the effect of LET on G_H, G_{OH}, $G_{e^-_{aq}}$, G_{H_2}, and $G_{H_2O_2}$ is still not definitely established. A reason for this is the rather limited number of systems which were studied systematically; most available data concern 0.4 M H_2SO_4 solutions of ferrous sulfate or ceric sulfate. Furthermore, the information provided by these experiments is extremely scarce concerning the dependence of measured $G(P)$ values on a number of factors, which are known to be of great importance for a reliable determination of G_R and G_M values (Sections I and II). Before 1960, when most of these investigations were carried out, the importance of true initial yield or of the scavenger reactivity was not so evident as it is today, after considerable efforts have been made to determine G_R and G_M in γ-radiolysis. It is to be hoped that after the great progress made in the field of radiolysis induced by radiations of low LET (γ rays and fast electrons), the investigators will resume the study of this almost forgotten subject. Meanwhile, we have to use the general picture given by Allen in his book [1], and by Bednář [89] and Pucheault [5] in their review articles.

Figure 5.5 represents the dependence of primary free-radical and molecular yields on radiation LET [1]. The data concern acid medium (0.4 M sulfuric acid) and were derived from various studies of ferrous ion oxidations and ceric ion reductions. In the calculation of primary yields, these were assumed to refer to the same species as in the case of γ-radiolysis in acid medium. It was also assumed that in the case of high-LET radiation the intraspur reaction

$$H_2O_2 + OH \longrightarrow HO_2 + H_2O \qquad (43)$$

cannot be disregarded and that the equation of material balance is correspondingly completed [Eq. (3a)]. It should be noted that for G_{HO_2} Allen [1] has assumed that the ratio G_{HO_2}/G_{H_2} increases with increasing LET as rapidly as does the ratio G_{H_2}/G_{red}, where $G_{red} = G_H + G_{e^-_{aq}}$. The results of a study [88], however, do not justify this assumption; it was found that G_{HO_2}/G_{H_2}

VI. EFFECT OF LINEAR ENERGY TRANSFER

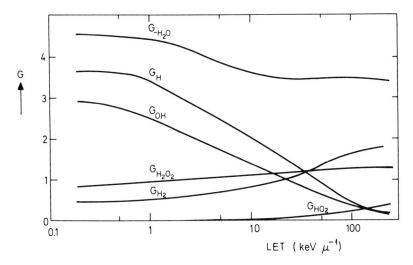

Fig. 5.5. Dependence of primary radical and molecular yields of water radiolysis on LET, derived from measurements on various systems in 0.4 M H_2SO_4. (After Allen [*1*].)

remains practically constant (0.06) for a rise in G_{H_2}/G_{red} from about 0.13 to 1.6.

The general picture presented in Fig. 5.5 can be completed with the results of only a few subsequent studies. Fricke dosimeter solutions were exposed to the α rays from ^{210}Po whose energy was lowered by means of mica absorbers [*90*]. It was shown that the value of $G(Fe^{3+})$ is decreased from 4.79 to 4.02 as the energy of α particles is decreased from 4.5 to 2.5 MeV. Thereafter, as the energy of α particles falls to 0.06 MeV, the yield of oxidation of ferrous ions again increases up to a value of 7.10. Vladimirova [*91a*] summarizes various data concerning the α-radiolysis of 0.4 M H_2SO_4 solutions in the presence and absence of $FeSO_4$ and $Ce(SO_4)_2$. The internal source of radiations was dissolved polonium. This analysis points out that the yield of water decomposition is 3.35 in such cases. This value is slightly lower than those given earlier by different authors, which varied between 3.5 and 3.7. All primary radical and molecular yields were determined in experiments with cyclotron accelerated protons [*91b*]. The oxidation of ferrous ion and the reduction of ceric ion (in the presence and absence of thallous ion) were systematically studied. It was found that as the proton energy decreases from 22 to 10.5 MeV, G_{red} decreases from 2.97 to 2.63 and G_{OH} decreases from 1.88 to 1.52. Surprisingly, under the same conditions molecular products remain unchanged: $G_{H_2O_2} = 0.92$ and $G_{H_2} = 0.39$. The stable products of radiolysis of an aqueous solution of formic acid (1 × 10^{-2} M), saturated with oxygen and containing 5 × 10^{-4} M sulfuric acid, have also been measured at various

LET [92]. Alpha rays of 3.4 MeV, as well as protons, deuterons, and helions, accelerated by a 60-inch cyclotron, were used for the irradiations. The energies varied from 20.9 MeV for deuterons (to which the lowest LET corresponded) down to 8.35 MeV for helions, with which the highest LET was attained. The reaction scheme of the radiolysis of formic acid in the presence of oxygen was considered in Sections III and IV. The same relations were used here for the calculation of the primary yields of the stable products observed (CO_2, H_2, H_2O_2, and for oxygen consumption). Detailed data show that G_M increases, whereas G_R and G_{-H_2O} decrease with increasing LET. The authors, however, drew attention to the fact that the yield of total water decomposition remains unchanged if one takes into account the yield of water reformation, favored under these conditions: $G(-H_2O)_{tot} = G_{-H_2O} + G(-H_2O)_{reform} = 4.5 \pm 0.1$. The yield of water reformation was assumed to be equal to the sum of the yields of H_2 and H_2O_2: $G(-H_2O)_{reform} = G_{H_2} + G_{H_2O_2}$. In studying the radiolytic oxidation of ferrous ions as well as the reduction of ceric ions under the action of α particles, Collinson et al. [93] considered in more detail yields referring to different effective segments of the α particle track in solution. The same authors [94] determined radical and molecular yields from measurements in solutions containing tritium as the internal radiation source (β-ray energy, 5.7 keV). They also carried out the measurements in heavy water solutions [95] with the same systems and radiation as in the preceding two works. Since the value of LET of tritium β rays lies between that of ^{60}Co γ-rays and ^{210}Po α-particles, it is interesting to compare G_R and G_M for light and heavy water derived from these experiments (Table 5.5). It can be calculated that the ratios of the radiation-chemical yields of the primary products in H_2O and D_2O have a trend similar to that reported for γ rays in Table 5.3.

The effects of reactivity of various substances on the yields of G_{H_2} were studied in irradiations with ^{60}Co γ-rays as well as with 18.9-MeV deuterons, mixed pile radiation, or fission fragments [96]. The results reported show that

TABLE 5.5

Yields of the Primary Radical and Molecular Products of H_2O and D_2O (0.1 N sulfuric acid) Radiolysis, Induced by Tritium β Particles [94, 95]

Primary radical products	G	Primary molecular products	G
$H + e_{aq}^-$	2.91	H_2	0.53
$D + e_{aq}^-$	3.34	D_2	0.27
OH	2.0	H_2O_2	0.97
OD	2.26	D_2O_2	0.81
HO_2	0	H_2O (dec)	3.97
DO_2	0	D_2O (dec)	3.89

also in the case of high-LET radiations the effect of reactivity is important. Aerated solutions of ferrous sulfate (10^{-2} M) in 0.4 M sulfuric acid were irradiated with C^{6+} ions, of an energy of 120 MeV, produced in a cyclotron [97]. For ions whose energy in solution was 102 MeV, $G(Fe^{3+})$ was found to be 4.94. The measurements were also carried out at the lower energies, and this technique seems very suitable for studying the effects of particles whose LET is 250 keV μ^{-1}, or even higher. The reduction of ceric ions by fragments formed in the nuclear reaction $^{10}B(n, \alpha)^7Li$ was studied in pile experiments by Sigli and Pucheault [98]. Appleby and Schwarz [88] derived the primary yields from nitrous oxide solutions containing ethanol, sodium deuteroformate, or potassium nitrite. Heavy ions were accelerated by a 60 inch cyclotron: 18-MeV deuterons, 32- and 12-MeV helium ions. Corresponding irradiations were also performed with ^{60}Co γ-rays. It was found that the hydrogen atom yield does not decrease as rapidly as the hydrated electron yield: G_H falls from 0.61 to 0.23 and $G_{e^-_{aq}}$ falls from 2.70 to 0.42 for ^{60}Co γ-rays and for 12-MeV He^{2+} ions, respectively. This finding is taken as an argument against direct production of hydrogen atoms by the action of the radiation on water as the sole source. Another finding is that for helium particles used in this study $G_{H_2} \sim G_{H_2O_2}$, also indicating that $G_{red} \sim G_{OH}$. Primary hydrogen peroxide yields were not measured, and the data obtained in a previous work with air-saturated KBr aqueous solutions [26] were used in the primary yield calculations.

VII. Effect of Dose Rate

The increase of dose rate should, in a way, have an effect similar to that observed in the case of the increase of LET, since in both cases an increase in density of primary events is involved. This means that a favoring of recombination reactions and an increase in G_M, with a simultaneous decrease in G_R, are to be expected. The yields of stable reaction products should increase or decrease, depending on whether these are formed by reactions with molecular or radical primary products.

The dose rate effect was studied by Keene [99], Rotblat and Sutton [100], Pikaev et al. [101–105], Anderson [106], Thomas and Hart [107], and Schwarz [40]. The irradiations were carried out with accelerated electrons, and the yield of oxidation of ferrous ions in 0.4 M sulfuric acid was often measured. Figure 5.6 shows this effect [108]. As can be seen, the $G(Fe^{3+})$ values decrease with increasing dose rate. The decrease is noticeable after about 10^8 rad sec^{-1}, but there are considerable differences as to the amount of lowering at a given dose rate. The cause of this should be sought in experimental conditions.

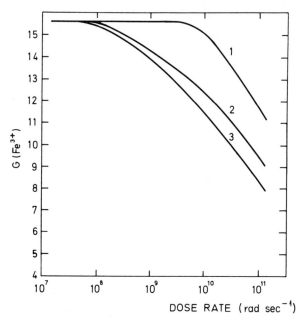

Fig. 5.6. Dependence of ferric ion yields on the dose rate of high-energy electrons, measured in aerated ferrous sulfate solutions in 0.4 M H_2SO_4: 1×10^{-2} M Fe^{2+} plus 1.2×10^{-3} M O_2 (1); aerated 1×10^{-3} M Fe^{2+} (2); aerated 1×10^{-3} M Fe^{2+} $+ 1 \times 10^{-3}$ M NaCl (3). (After Fricke and Hart [108].)

The yields of H_2 and H_2O_2 were measured in deaerated aqueous solutions of 0.4 M H_2SO_4 [109]. The solutions were irradiated with 1.4-μsec pulses of 15-MeV electrons from an accelerator. The dose rate in solution during the pulse was 2×10^{23} eV ml^{-1} sec^{-1}. It is interesting to note that the two measured yields were found to have practically the same value. In 0.4 M H_2SO_4 solution $G(H_2) \sim G(H_2O_2) \sim 1.2$. It was also found that the increase of reactivity toward reducing species leads to the decrease of the molecular hydrogen yields. Figure 5.7 shows how the increase of concentration of H_2O_2 lowers $G(H_2)$ in 0.4 M H_2SO_4 deaerated solutions. For comparison, corresponding data on the effect of reactivity in the case of γ-radiolysis are also plotted on the diagram. We see that an increase in concentration of hydrogen peroxide from 10^{-6} M to about 10^{-2} M causes a drastic decrease in $G(H_2)$ in the case of irradiation at a high dose rate, while at a low dose rate there is no change in $G(H_2)$ values. A further increase in reactivity toward primary reducing species leads to a decrease in measured molecular hydrogen yields, but this seems to be independent of the dose rate under these working conditions. Thomas [110] also found that at higher dose rates the scavengers are more effective in decreasing molecular hydrogen yield. He studied acidic

VIII. EFFECT OF TEMPERATURE 157

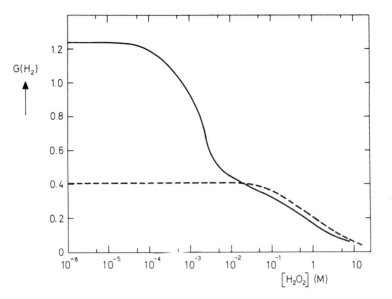

Fig. 5.7. Effect of reactivity toward e_{aq}^{-} on the yield of molecular hydrogen at high dose rate (2×10^{23} eV ml^{-1} sec^{-1}). The dashed line represents the corresponding dependence at low dose rates of γ-radiolysis. (After Anderson and Hart [109].)

aqueous solutions containing hydrogen peroxide, oxygen, or cupric ion. The irradiations were performed with single pulses of 13 MeV electrons, over the dose rate range 10^{16} to 10^{19} eV ml^{-1} in a 1.4-μsec pulse.

Willis et al. [111] have reported experimental and calculated yields of molecular hydrogen measured in water at dose rates between 10^{25} and 10^{28} eV ml^{-1} sec^{-1}. Particular attention was paid to the dosimetry, as well as to the dose rate monitoring. It was found that $G(H_2)$ in pure water decreases from about 1 to 0.8 as the dose rate rises from 9×10^{25} to 4.8×10^{27} eV ml^{-1} sec^{-1}. The increase in acidity (pH change from 7 to 1.1) causes an increase in $G(H_2)$ from 0.8 to about 1.1 at the highest dose rate.

Different aspects of water radiolysis at very high dose rates of electrons were systematically studied by Pikaev et al. [112–114].

VIII. Effect of Temperature

The increase of temperature might lead to a broader distribution of primary species and thereby to an increase in G_R and a decrease in G_M. Hochanadel and Ghormley experimentally checked this assumption [115]. Varying the temperature from 2 to 65°C, the authors measured the yields of a number of reactions whose mechanisms are reliably established in γ-radiolysis: oxidation of ferrous ions in the presence and absence of oxygen, and reduction of ceric ions and generation of hydrogen peroxide in aerated water and aerated and deaerated solutions of KBr. The solvent in all the cases was 0.4 M sulfuric acid. The temperature coefficients determined ranged from 0.06 to 0.20% per degree. They were positive for the ferrous ion oxidation and negative for the other processes. Table 5.6 shows the observed effect of temperature on the yields of primary species calculated from experimental results, assuming the reaction schemes given in other papers. We see that the trend of changes caused by the increase of temperature is indeed as expected—the rise of G_R and the fall of G_M. They are very small, however, and it is understandable why usual variations of room temperature have no appreciable effect on measured yields of primary products.

TABLE 5.6.

DEPENDENCE OF PRIMARY RADICAL AND MOLECULAR YIELDS OF WATER γ-RADIOLYSIS ON TEMPERATURE, DERIVED FROM VARIOUS SYSTEMS (0.4 M H_2SO_4).[a]

Primary species	G			Temperature coefficient, % °C^{-1}
	2°C	23°C	65°C	
$H + e_{aq}^-$	3.59	3.67	3.82	$+0.10 \pm 0.03$
OH	2.80	2.91	3.13	$+0.18 \pm 0.04$
H_2	0.38	0.37	0.36	-0.06 ± 0.03
H_2O_2	0.78	0.75	0.70	-0.15 ± 0.03
$-H_2O$	4.35	4.41	4.54	$+0.07 \pm 0.03$

[a] After Hochanadel and Ghormley [115].

The effect of temperature on reaction rate constants is also not large; these increase only by a few tenths of a percent per degree. Nevertheless, it has been successfully used for determination of some activation energies on the basis of the Arrhenius law:

$$\ln k = -E/RT + \ln A$$

where

- E is the activation energy,
- k is the rate constant at the given temperature,
- T is the absolute temperature,
- R is the gas constant,
- A is the frequency factor.

It has been experimentally shown [116] that different rate constants increase with rising temperature and that, within the limits of experimental error ($\pm 15\%$), they fit a graph plotted according to the Arrhenius law. The authors observed a number of reactions of the hydrated electron in light and heavy water, whose rate constants range from 10^6 to 10^9 M^{-1} sec^{-1}, and they obtained values between 2.8 and 4.0 kcal mole^{-1} for activation energies.

Varying the temperature between 20 and 80°C, Čerček and Ebert [117,118] determined activation energies for the reaction of e_{aq}^- with NO_2^- and H_3O^+ to be 2.6 and 1.7 kcal mole^{-1}, respectively. These values are substantially lower than the activation energy for diffusion in water, and the authors concluded that the mechanism by which e_{aq}^- moves in water considerably differs from normal diffusion. It was also found [119] that the reactions of e_{aq}^- do not have a constant energy of activation, the reproducibility of these measurements was within 10%.

IX. Effect of Pressure

Considerable structural changes occur in water subjected to pressures substantially higher than atmospheric. Already at 1.5–2.0 kbar the dissociation constant is known to be four times as large as that at atmospheric pressure. Such a pressure also makes water behave, at a temperature of 20°C, as a "simple," unassociated liquid. The viscosity at 20°C increases from 10 to 15 mP as the pressure increases to 7 kbar.

Similar to the preceding expression for the dependence of the reaction rate constant on temperature, we have the following relation for pressure [120]:

$$\ln k = -\frac{\Delta V^{\ddagger}}{RT} P + \ln k_0$$

Here, k and k_0 are reaction rate constants measured at pressures P and zero, respectively (k_0 being very close to what is measured at atmospheric pressure), R is the gas constant (1.986 cal grad^{-1} mole^{-1}), and T is the absolute temperature. The symbol $\Delta V^{\ddagger}$ denotes the activation volume which is expressed in milliliters per mole. It is a characteristic of the reaction, just as the entropy of activation; its physical meaning is the difference between the volumes of reactants and the activation complex. For graphical representation, use is more often made of an expression of the type $y = -ax$. The activation volume may have a positive sign, which denotes that $\Delta V^{\ddagger}$ increases with increasing pressure, or a negative sign. It is assumed that the latter is the case of bimolecular processes where the structural changes of reactants in the activation complex are such that $\Delta V^{\ddagger}$ is negative, in contrast to the case of monomolecular processes. However, the real situation in solutions is much more complex, particularly when the ions or strong dipoles are in question; then rearrangement of the molecules of solvent has a decisive effect on the activation volume.

Hentz et al. [121–127] showed by a series of very carefully performed experiments that an increase in atmospheric pressure to 6.4 kbar has no effect on the yields of primary species, although it affects the yields of some stable products of the reactions studied. The investigated solutions were placed in a glass vessel, which was put in a shielding reaction vessel made of steel during irradiation. In the gap between the latter and the glass vessel, n-hexan was used as the hydrostatic fluid. A high hydrostatic pressure, achieved by means of a special device, was exerted on a piston which entered the glass vessel with the solution investigated. At the highest pressures the decrease in volume of the solution amounted to about 15%.

Aerated and deaerated solutions of ferrous sulfate were irradiated in the presence of sulfuric acid [121]. The yields of oxidation of ferrous ions, oxygen consumption, and molecular hydrogen production were measured. Thus, for example, in solutions containing oxygen (2.4×10^{-4} M) and sulfuric acid (0.4 M), an increase in pressure from 0 to 6.34 kbar leads to an increase in $G(H_2)$ from 0.44 to 0.80, with a simultaneous decrease in $G(Fe^{3+})$ and $G(-O_2)$ from 15.6 and 3.66 to 14.68 and 3.24, respectively. Analysis of the experimental data, however, shows that the following sums of measured yields are independent of pressure:

$$G(H_2) + G(-O_2) = 4.1$$

$$G(Fe^{3+}) + 2G(H_2) = 16.4$$

Kinetic analysis of the process shows that the first expression is equal to $G_H + G_{H_2}$ and the second to $4(G_H + G_{H_2})$. Hence, it follows that there is no effect of pressure on the yields of primary reducing species, hydrogen atoms, and molecular hydrogen.

Similar experiments on reduction of ceric ions in solutions of 0.4 M sulfuric acid provided supplementary data necessary for calculating oxidizing primary yields [122]. The yields of reduction of ceric ions in deaerated solutions and the yields of production of H_2 and O_2 were measured. The conclusion is that there is no effect of pressure on the primary yields of free radicals and molecular products in acid medium.

A number of similar measurements were carried out on different systems at neutral pH [123]. The experimental conditions were the same as in the experiments in acid medium described above. It was also established that an increase in pressure to 6.34 kbar does not affect the yields of primary molecular and radical products.

While no effect of pressure on the primary yields of water radiolysis could be proved, the rate constants of some of the reactions studied were found to be appreciably affected by change in pressure. Thus, for example, data on the

IX. EFFECT OF PRESSURE

increase in $G(H_2)$ and decrease in $G(-O_2)$ with increasing pressure pointed out [121] the effect of pressure on the competition:

$$H + O_2 \longrightarrow HO_2 \qquad [5]$$

$$H + H_3O^+ + Fe^{2+} \longrightarrow Fe^{3+} + H_2 \qquad (44)$$

The ratio k_5/k_{44} falls from 1000 at atmospheric pressure to only 80 at 6.34 kbar. From these measurements, $\Delta V_5^{\ddagger} - \Delta V_{44}^{\ddagger}$ is calculated to be 11 ml mole^{-1}. The dependence of water viscosity on pressure allows one to estimate $\Delta V_5^{\ddagger}$ to be 2 ml mole^{-1} and to calculate the activation volume for reaction (44) to be $\Delta V_{44}^{\ddagger} = -9$ ml mole^{-1}. The authors consider this value to be consistent with the assumption that reaction (44) first produces FeH^{2+}, which then reacts with H_3O^+ to give the stable products of reaction (44).

In the radiolysis of aqueous solutions of bicarbonate, also containing isopropanol, an increase in pressure to 6.4 kbar leads to an appreciable increase in the measured yield of molecular hydrogen [124]. This was explained by the effect of pressure on the competition:

$$e_{aq}^- + CO_2 \longrightarrow CO_2^- \qquad (45)$$

$$e_{aq}^- + HCO_3^- \longrightarrow H + CO_3^{2-} \qquad (46)$$

An increase in pressure favors reaction (46) and thereby the production of H_2 in the reaction of the hydrogen atom with alcohol (H + HR → H_2 + R) which follows reaction (46). The authors made use of these results to calculate the activation volume for the electron. Proceeding from the fact that the absolute activation volume for the proton, $\bar{V}_0(H^+)$, is equal to -5.4 ml mole^{-1}, they calculated the absolute activation volume for the hydrated electron, $\bar{V}_0(e_{aq}^-)$, to be within the range -5.5 to -1.1 ml mole^{-1}.

Hentz et al. [125] have also studied the effect of pressure on the formation of H_2 in the γ-radiolysis of deaerated solutions of glucose in the presence of H_2O_2 or KNO_3. The increase of pressure also leads to the increase of $G(H_2)$. It is assumed that pressure affects the competition as given by Eqs. (47) and (48),

$$e_{aq}^- + NO_3^- \text{ (or } H_2O_2) \not\longrightarrow H \text{ or } H_2 \qquad (47)$$

$$e_{aq}^- + H_2O \longrightarrow H + OH^- \qquad (48)$$

by favoring reaction (48) and, via the reaction H + HR → H_2 + R which follows it, the production of H_2. From the experimental data, the activation volume for the hydrated electron is estimated to be within the range -5.5 to -1.1 ml mole^{-1}, assuming $\bar{V}_0(H^+)$ to be -5.4 ml mole^{-1}, as in the preceding case. The authors think that this estimate is in good agreement with the data obtained from the radiolysis of bicarbonate.

References

1. A. O. Allen, "The Radiation Chemistry of Water and Aqueous Solutions," Van Nostrand-Reinhold, Princeton, New Jersey, 1961.
2. M. Haissinsky, Rendements radiolytiques primaires en solution aqueuse, neutre ou alcaline. *In* "Actions chimiques et biologiques des radiations" (M. Haissinsky, ed.), Vol. 11, p. 133. Masson, Paris, 1967.
3. E. Hayon, Primary Radicals Yields in the Radiation Chemistry of Water and Aqueous Solutions. *In* "Radiation Chemistry of Aqueous Systems" (G. Stein, ed.), p. 157. Wiley (Interscience), New York, 1968.
4. G. Czapski, Radical Yields as a Function of pH and Scavenger Concentration. *Advan. Chem. Ser.* **81**, 106 (1968).
5. J. Pucheault, Action des rayons alpha sur les solutions aqueuses. *In* "Actions chimiques et biologiques des radiations" (M. Haissinsky, ed.), Vol. 5, p. 33. Masson, Paris, 1961.
6. M. Anbar, Water and Aqueous Solutions. *In* "Fundamental Processes in Radiation Chemistry" (P. Ausloos, ed.), p. 651. Wiley (Interscience), New York, 1968.
7. G. V. Buxton, Primary Radical and Molecular Yields in Aqueous Solution, The Effect of pH and Solute Concentration. *Radiat. Res. Rev.* **1**, 209 (1968).
8. O. D. Bonner and G. B. Woolsey, The Effect of Solutes and Temperatures on the Structure of Water. *J. Phys. Chem.* **72**, 899 (1968).
9. R. A. Horne, The Structure of Water and Aqueous Solutions. *Surv. Progr. Chem.* **4**, 1 (1968).
10. D. Eisenberg and W. Kauzmann, "The Structure and Properties of Water." Oxford Univ. Press (Clarendon), London and New York, 1969.
11. A. O. Allen, The Yields of Free H and OH in the Irradiation of Water. *Radiat. Res.* **1**, 85 (1954).
12. A. O. Allen, A Survey of Recent American Research in the Radiation Chemistry of Aqueous Solutions, *Proc. Int. Conf. Peaceful Uses At. Energy, 2nd, Geneva, 1958* **7**, P/738, 513 (1958).
13. C. J. Hochanadel, Radiation Chemistry of Water. *In* "Comparative Effects of Radiation" (M. Burton, J. S. Kirby-Smith, and J. L. Magee, eds.), Chapter 8, p. 151. Wiley, New York, 1960.
14. I. G. Draganić, M. T. Nenadović, and Z. D. Draganić, Radiolysis of $HCOOH + O_2$ at pH 1.3–13 and the Yields of Primary Products in γ-Radiolysis of Water. *J. Phys. Chem.* **73**, 2564 (1969).
15. E. J. Hart, γ-Ray Induced Oxidation of Aqueous Formic Acid-Oxygen Solutions. Effect of pH. *J. Amer. Chem. Soc.* **76**, 4198 (1954).
16. J. C. Russel and G. R. Freeman, Yield of Solvated Electrons in the Gamma Radiolysis of Water + 10% Ethanol: Nonhomogeneous Kinetics of Electron Scavenging in Water. *J. Chem. Phys.* **48**, 90 (1968).
17. F. S. Dainton and S. R. Logan, Radiolysis of Aqueous Solutions Containing Nitrite Ions and Nitrous Oxide. *Trans. Faraday Soc.* **61**, 715 (1965).
18. T. J. Sworski, Yields of Hydrogen Peroxide in the Decomposition of Water by Cobalt γ-Radiation. I. Effect of Bromide Ion. *J. Amer. Chem. Soc.* **76**, 4687 (1954).
19. T. J. Sworski, Yields of Hydrogen Peroxide in the Decomposition of Water by Cobalt γ-Radiation. II. Effect of Chloride Ion. *Radiat. Res.* **2**, 26 (1955).
20. H. A. Schwarz, The effect of Solutes on Molecular Yields in the Radiolysis of Aqueous Solutions. *J. Amer. Chem. Soc.* **77**, 4960 (1955).

REFERENCES

21. A. O. Allen and R. A. Holroyd, Peroxide Yield in the γ-Irradiation of Air-Saturated Water. *J. Amer. Chem. Soc.* **77**, 5852 (1955).
22. J. A. Ghormley and C. J. Hochanadel, The Effect of Hydrogen Peroxide and Other Solutes on the Yield of Hydrogen in the Decomposition of Water by γ-Rays. *Radiat. Res.* **3**, 227 (1955).
23. T. J. Sworski, Mechanism for the Reduction of Ceric Ion by Thallous Ion Induced by Cobalt-60 Gamma Radiation. *Radiat. Res.* **4**, 483 (1956).
24. H. A. Schwarz and A. J. Salzman, The Radiation Chemistry of Aqueous Nitrite Solutions: The Hydrogen Peroxide Yield. *Radiat. Res.* **9**, 502 (1958).
25. H. A. Mahlman, Ceric Reduction and the Radiolytic Hydrogen Yield. *J. Amer. Chem. Soc.* **81**, 3203 (1959).
26. H. A. Schwarz, J. M. Caffrey and G. Scholes, Radiolysis of Neutral Water by Cyclotron Produced Deuterons and Helium Ions. *J. Amer. Chem. Soc.* **81**, 1801 (1959).
27. C. N. Trumbore and E. J. Hart, α-Ray Oxidation of Ferrous Sulfate in 0.4 M Sulfuric Acid Solutions. The Effect of 0 to 0.4 M Oxygen. *J. Phys. Chem.* **63**, 867 (1959).
28a. H. A. Mahlman, Activity Concept in Radiation Chemistry, *J. Chem. Phys.* **31**, 993 (1959).
28b. H. A. Mahlman, Hydrogen Formation in the Radiation Chemistry of Water. *J. Chem. Phys.* **32**, 601 (1960).
29. G. Hughes and C. Willis, Scavenger Studies in the Radiolysis of Aqueous Ferricyanide Solutions at High pH. *Discuss. Faraday Soc.* **36**, 223 (1963).
30. A. Rafi and H. C. Sutton, Radiolysis of Aerated Solutions of Potassium Bromide. *Trans. Faraday Soc.* **61**, 877 (1965).
31a. Z. D. Draganić, I. G. Draganić, and M. M. Kosanić, Radiation Chemistry of Oxalate Solutions in the Presence of Oxygen over a Wide Range of Acidities. *J. Phys. Chem.* **68**, 2085 (1964).
31b. Z. D. Draganić, I. G. Draganić, and M. M. Kosanić, Radiolysis of Oxalate Alkaline Solutions in the Presence of Oxygen. *J. Phys. Chem.* **70**, 1418 (1966).
32. E. Hayon and M. Moreau, Reaction Mechanism Leading to the Formation of Molecular Hydrogen in the Radiation Chemistry of Water. *J. Phys. Chem,* **69**, 4058 (1965).
33. H. A. Schwarz, *Proc. Informal Conf. Radiat. Chem. Water, 5th* Univ. of Notre Dame. Radiation Laboratory AEC Doc. No. COO-38-519, p. 51. Notre Dame, Indiana. 1966.
34. M. Chouraqui and J. Sutton, Origin of Primary Hydrogen Atom Yield in Radiolysis of Aqueous Solutions. *Trans. Faraday Soc.* **62**, 2111 (1966).
35. Z. D. Draganić and I. G. Draganić, On the Origin of Primary Hydrogen Peroxide Yield in the γ-Radiolysis of Water. *J. Phys. Chem.* **73**, 2571 (1969).
36a. E. Hayon, Solute Scavenging Effects in Regions of High Radical Concentration Produced in Radiation Chemistry. *Trans. Faraday Soc.* **61**, 723 (1965).
36b. E. Hayon, Radical and Molecular Yields in the Radiolysis of Alkaline Aqueous Solutions, *Trans. Faraday Soc.*, **61**, 734 (1965).
37. D. A. Flanders and H. Fricke, Application of a High–Speed Electronic Computer in Diffusion Kinetics. *J. Chem. Phys.* **28**, 1126 (1958).
38. G. V. Buxton and F. S. Dainton, Radical and Molecular Yields in the γ-Radiolysis of Water. II. The Potassium Iodide-Nitrous Oxide System in the pH Range 0 to 14. *Proc. Roy. Soc. (London)* A **287**, 427 (1965).
39. C. E. Burchill, F. S. Dainton, and D. Smithies, Radical and Molecular Product Yields in X-Irradiated Alkaline Aqueous Solutions. *Trans. Faraday Soc.* **63**, 932 (1967).

40. H. A. Schwarz, Intensity Effects, Pulsed-Beam Effects, and the Current Status of Diffusion Kinetics. *Radiat. Res. Suppl.* **4**, 89 (1964).
41. J. Pucheault and C. Ferradini, Actions chimiques des radiations ionisantes sur les solutions aqueuses de vanadium. IV. Influence des ions Cl⁻ sur la reduction des solutions acides de vanadium pentavalent par les rayons γ. *J. Chim. Phys.* **58**, 606 (1961).
42. C. H. Cheek and V. J. Linnenbom, The Radiation Chemistry of Alkaline Hydrobromite Solutions. *J. Phys. Chem.* **67**, 1856 (1963).
43. M. Daniels and E. E. Wigg, Radiation Chemistry of the Nitrate System. II. Scavenging and pH Effects in the Cobalt-60 γ-Radiolysis of Concentrated Sodium Nitrate Solutions. *J. Phys. Chem.* **73**, 3703 (1969).
44. M. Daniels, Radiation Chemistry of the Aqueous Nitrate System. III. Pulse Electron Radiolysis of Concentrated Sodium Nitrate Solutions. *J. Phys. Chem.* **73**, 3710 (1969).
45. G. E. Adams and R. L. Willson, Pulse Radiolysis Studies on the Oxidation of Organic Radicals in Aqueous Solution. *Trans. Faraday Soc.* **65**, 2981 (1969).
46. G. E. Adams and E. J. Hart, Radiolysis and Photolysis of Aqueous Formic Acid. Carbon Monoxide Formation. *J. Amer. Chem. Soc.* **84**, 3994 (1962).
47. E. J. Hart, Mechanism of the γ-Ray Induced Oxidation of Formic Acid in Aqueous Solutions. *J. Amer. Chem. Soc.* **73**, 68 (1951).
48. J. P. Keene, Y. Raef, and A. J. Swallow, Pulse Radiolysis Studies of Carboxyl and Related Radicals. *In* "Pulse Radiolysis" (M. Ebert, J. P. Keene, A. J. Swallow, and J. H. Baxendale, eds.), p. 99. Academic Press, New York, 1965.
49. Z. D. Draganić and M. T. Nenadović, Use of ^{14}C in Studying the Reaction of Oxalic Acid Re-formation During Gamma Radiolysis of Aqueous Oxalate Solutions. *Int. J. Appl. Radiat. Isotopes* **17**, 319 (1966).
50. G. Czapski and L. M. Dorfman, Pulse Radiolysis Studies. V. Transient Spectra and Rate Constants in Oxygenated Aqueous Solutions. *J. Phys. Chem.* **68**, 1169 (1964).
51. J. Rabani and S. O. Nielsen, Absorption Spectrum and Decay Kinetics of O_2^- and HO_2 in Aqueous Solutions by Pulse Radiolysis. *J. Phys. Chem.* **73**, 3736 (1969).
52. E. J. Hart, J. K. Thomas, and S. Gordon. A Review of the Radiation Chemistry of Single-Carbon Compounds and Some Reactions of the Hydrated Electron in Aqueous Solutions. *Radiat. Res. Suppl.* **4**, 74 (1964).
53. J. L. Weeks and J. Rabani, The Pulse Radiolysis of Deaerated Aqueous Carbonate Solutions. I. Transient Optical Spectrum and Mechanism. II. pK for OH Radicals. *J. Phys. Chem.* **70**, 2100 (1966).
54. E. M. Fielden and E. J. Hart, Primary Radical Yields in Pulse-Irradiated Alkaline Aqueous Solution. *Radiat. Res.* **32**, 564 (1967).
55. E. M. Fielden and E. J. Hart, Primary Radical Yields and Some Rate Constants in Heavy Water. *Radiat. Res.* **33**, 426 (1968).
56. A. Kuppermann, Diffusion Model of the Radiation Chemistry of Aqueous Solutions. *In* "Radiation Research 1966" (G. Silini, ed.) p. 212. North-Holland Publ., Amsterdam, 1967.
57. B. H. J. Bielski and A. O. Allen, The Radiolytic Yield of Reducing Radicals in Neutral Aqueous Solutions, *Int. J. Radiat. Phys. Chem.* **1**, 153 (1969).
58. A. O. Allen and H. A. Schwarz, Decomposition of Water under High Energy Radiation. *Proc. Int. Conf. Peaceful Uses At. Energy, 2nd Geneva, 1958* **29**, P/1403, 30 (1958).
59. M. Haissinsky, Radiolyse γ de solutions alcalines et neutres. I. Solutions alcalines de phosphite de sodium. II. Effet du pH sur le rendements. *J. Chim. Phys.* **62**, 1141 (1965).

60. C. J. Hochanadel and R. Casey, The Yield of Reducing Radicals in the Gamma Radiolysis of Water Using Carbon Monoxide and Oxygen as Scavengers. *Radiat. Res.* **25**, 198 (1965).
61. T. Sawai, Radical Yields in the Radiolysis of Neutral Aqueous Solutions. *Bull. Chem. Soc. (Japan)* **39**, 955 (1966).
62. J. H. Baxendale and A. A. Khan, The Pulse Radiolysis of p-Nitrosodimethylaniline in Aqueous Solution. *Int. J. Radiat. Phys. Chem.* **1**, 11 (1969).
63. S. Nehari and J. Rabani, The Reaction of H Atoms with OH^- in the Radiation Chemistry of Aqueous Solutions. *J. Phys. Chem.* **67**, 1609 (1963).
64. E. Hayon and M. Moreau, Réactivité des atomes H avec quelques composés organiques et minéraux en solutions aqueuses. *J. Chim. Phys.* **62**, 391 (1965).
65. H. A. Mahlman, Radiolysis of Nitrous Oxide Saturated Solutions: Effect of Sodium Nitrate, 2-Propanol, and Sodium Formate. *J. Phys. Chem.* **70**, 3983 (1966).
66. H. A. Schwarz, J. P. Losee, and A. O. Allen, Hydrogen Yields in the Radiolysis of Aqueous Solutions. *J. Amer. Chem. Soc.* **76**, 4693 (1954).
67. H. A. Mahlman and J. W. Boyle, Primary Cobalt-60 Radiolysis Yields in Heavy Water, *J. Amer. Chem. Soc.* **80**, 773 (1958).
68. J. D. Backhurst, G. R. A. Johnson, G. Scholes, and J. Weiss, Determination of the Yields of Molecular Hydrogen Peroxide in the Radiolysis of Water Labeled with Oxygen-18. *Nature* **183**, 176 (1959).
69. J. W. Boyle, The Decomposition of Aqueous Sulfuric Acid Solutions by Cobalt Gamma Rays. II. Yields of Solvent Decomposition and Reducing Radicals from Fe (II) Solutions in 0.4 to 18 M Acid. *Radiat. Res.* **17**, 450 (1962).
70. G. Czapski and E. Peled, On the pH-Dependence of $G_{reducing}$ in the Radiation Chemistry of Aqueous Solutions. *Israel J. Chem.* **6**, 421 (1968).
71. G. Czapski and B. J. Bielski, The Formation and Decay of H_2O_3 and HO_2 in Electron-Irradiated Aqueous Solutions. *J. Phys. Chem.* **67**, 2180 (1963).
72. B. L. Gall and L. M. Dorfman, Pulse Radiolysis. XV. Reactivity of the Oxide Radical Ion and of the Ozonide Ion in Aqueous Solution. *J. Amer. Chem. Soc.* **91**, 2199 (1969).
73. C. H. Cheek, V. J. Linnenbom, and J. W. Swinnerton, The Effect of Acidity on the Hydrogen Yield in Gamma-Irradiated Aqueous Solutions. *Radiat. Res.* **19**, 636 (1963).
74. C. Lifshitz and G. Stein, Isotope Effects in the Radiation Chemistry of Water. *Israel J. Chem.* **2**, 337 (1964).
75. M. Anbar and D. Meyerstein, Isotopically Substituted Water in the Investigation of the Primary Radiolytic Processes. *In* " Radiation Chemistry of Aqueous Systems " (G. Stein, ed.), p. 109. Wiley (Interscience), New York, 1968.
76. T. J. Hardwick, Measurements of the Recombination of H and OH in γ-Irradiated Aqueous Systems. *J. Chem. Phys.* **31**, 226 (1959).
77. D. A. Armstrong, E. Collinson, and F. S. Dainton, Radical and Molecular Yields in Solutions of Acrylamide in Light and Heavy Water. *Trans. Faraday Soc.* **55**, 1375 (1959).
78. G. Némethy and H. A. Scheraga, Structure of Water and Hydrophobic Bonding in Proteins. IV. The Thermodynamic Properties of Liquid Deuterium Oxide. *J. Chem. Phys.* **41**, 680 (1964).
79. C. M. Davis, Jr. and D. L. Bradley, Two-State Theory of the Structure of D_2O. *J. Chem Phys.* **45**, 2461 (1966).
80. R. L. Platzman, Superexcited States of Molecules, and the Primary Action of Ionizing Radiation. *Vortex* **23**, 372 (1962).

81. E. Hayon, Radiolysis of Heavy Water in the pD Range 0–14. *J. Phys. Chem.* **69**, 2628 (1965).
82. Z. D. Draganić, O. Mićić, and M. Nenadović, Radiolysis of Some Heavy Water Solutions at pD 1.3–13. *J. Phys. Chem.* **72**, 511 (1968).
83a. K. Coatsworth, E. Collinson, and F. S. Dainton, Radical and Molecular Yields in Acidified Deuterium Oxide. *Trans. Faraday Soc.* **56**, 1008 (1960).
83b. B. Bartoníček, J. Janovský, and J. Bednář, Localization of Energy in Radiolysis of Solutions. II. Yields of OH and OD Radicals in Mixtures of Light and Heavy Water. *Collect. Czech. Chem. Commun.* **30**, 1328 (1965).
84. K. D. Asmus and J. H. Fendler, The Use of Sulfur Hexafluoride to Determine $G(e_{D_2O}^-)$ and Relative Reaction Rate Constants in D_2O. *J. Phys. Chem.* **73**, 1583 (1969).
85. M. T. Nenadović, Z. D. Draganić, and I. G. Draganić, Radiolysis of Formic Acid-Oxygen at pD 1.3–13 and the Yields of Primary Products in γ-Radiolysis of Heavy Water. *In* "Proceedings of the Third Tihany Symposium on Radiation Chemistry" (J. Dobó and P. Hedvig, eds.), Akademiai Kiado, Budapest (submitted for publication).
86. E. Hart and M. Fielden, Reaction of the Deuterated Electron, e_d^-, with e_d^-, D, OD and D_2O. *J. Phys. Chem.* **72**, 577 (1968).
87. Report of the International Commission on Radiobiological Units and Measurements (ICRU), 1959, US Dep. of Commerce, Nat. Bur. of Std. Handbook 78.
88. A. Appleby and H. A. Schwarz, Radical and Molecular Yields in Water Irradiated by γ Rays and Heavy Ions. *J. Phys. Chem.* **73**, 1937 (1969).
89. J. Bednář, Radiační chemie vody a vodných roztoku anorganických látek. *Chem. Listy* **53**, 1083 (1959).
90. S. Gordon and E. J. Hart, Chemical Yields of Ionizing Radiations in Aqueous Solutions: Effect of Energy of Alpha Particles. *Radiat. Res.* **15**, 440 (1961).
91a. M. V. Vladimirova, Molecular and Radical Product Yields in Alpha-Radiolysis of $0.8\ N\ H_2SO_4$. *Advan. Chem. Ser.* **81**, 280 (1968).
91b. G. L. Kochanny, Jr., A Timnick, C. J. Hochanadel, and C. D. Goodman, Radiation Chemistry Studies of Water as Related to the Initial Linear Energy Transfer of 11-MeV to 23-MeV Protons. *Radiat. Res.* **19**, 462 (1963).
92. A. R. Anderson and E. J. Hart, Molecular Product and Free Radicals Yields in the Decompositions of Water by Protons, Deuterons, and Helium Ions. *Radiat. Res.* **14**, 689 (1961).
93. E. Collinson, F. S. Dainton, and J. Kroh, The Radiation Chemistry of Aqueous Solutions. I. The Effect of Changing Linear Energy Transfer along a Polonium α-Particle Track. *Proc. Roy. Soc. (London)* **A265**, 407 (1962).
94. E. Collinson, F. S. Dainton, and J. Kroh, The Radiation Chemistry of Aqueous Solutions. II. Radical and Molecular Yields for Tritium β-Particles. *Proc. Roy. Soc. (London)* **A265**, 422 (1962).
95. E. Collinson, F. S. Dainton and J. Kroh, The Radiation Chemistry of Aqueous Solutions. III. The Isotope Effect for Polonium α-Particles and Tritium β-Particles. *Proc. Roy. Soc. (London)* **A265**, 430 (1962).
96. T. J. Sworski, Kinetic Evidence that "Excited Water" is Precursor of Intraspur H_2 in the Radiolysis of Water, *Advan Chem. Ser.* **50**, 263 (1965).
97. R. H. Schuler, Radiation Chemical Studies with Heavy Ion Radiations. *J. Phys. Chem.* **71**, 3712 (1967).
98. P. Sigli and J. Pucheault, Radiolyses de solutions aqueuses par des rayonnements á ionisations denses. Radioréduction induite par la réaction nucléaire $^{10}B(n, \alpha)^7Li$ dans des solutions sulfuriques de Ce^{IV} en présence de Tl^I et de Ce^{III}. *J. Chim. Phys.* **65**, 1543 (1968).

99. J. P. Keene, The Oxidation of Ferrous Ammonium Sulfate Solutions by Electron Irradiation at High Dose Rates. *Radiat. Res.* **6**, 434 (1957).
100. J. Rotblat and H. C. Sutton, The Effect of High Dose Rates of Ionizing Radiations on Solutions of Iron and Cerium Salts. *Proc. Roy Soc.* (*London*) **A255**, 490 (1960).
101. A. K. Pikaev and P. Ya. Glazunov, Issledovanie radioliticheskogo okislenia dvukhvalentnogo zheleza pri moshchnosti dozy 10^{21} eV/ml. sek. *Izv. Akad. Nauk SSSR Otd. Khim. Nauk* 2244 (1959).
102. A. K. Pikaev and P. Ya. Glazunov, Vlianie kontsentratsii rastvora na radiatsionnyi vykhod okislenia dvukhvalentnogo zheleza pri vysokykh moshchnostiakh dozy obluchenia. *Izv. Akad. Nauk SSSR Otd. Khim. Nauk* 2063 (1960).
103. P. Ya. Glazunov and A. K. Pikaev, Issledovanie radioliticheskogo okislennia dvukhvalentnogo zheleza pri vysokykh moshchnostiakh dozy obluchenia. *Dokl. Akad. Nauk. SSSR* **130**, 1051 (1960).
104. A. K. Pikaev, P. Ya. Glazunov, and I. Spitsyn, Mekhanizm radioliticheskogo okisleniia dvukhvalentnogo zheleza v sernokyslikh vodnykh rastvorakh, soderzhashchikh kislorod, pri vysokikh moshchnostiakh pogloshchennoi dozy, *Dokl. Akad. Nauk SSSR* **150**, 1077 (1963).
105. A. K. Pikaev, P. Ya. Glazunov, Radioliz vodnykh rastvorov ferrosulfata pod deistviem detsimikrosekundnykh impulsov elektronov. *Dokl. Akad. Nauk SSSR* **154**, 1167 (1964).
106. A. R. Anderson, A Calorimetric Determination of the Oxidation Yield of the Fricke Dosimeter at High Dose Rates of the Electrons. *J. Phys. Chem.* **66**, 180 (1962).
107. J. K. Thomas and E. J. Hart, The Radiolysis of Aqueous Solutions at High Intensities. *Radiat. Res.* **17**, 408 (1962).
108. H. Fricke and E. J. Hart, Chemical Dosimetry. *In* "Radiation Dosimetry" (F. H. Attix and W. C. Roesh, eds.), Vol. 2, Academic Press, New York, 1966.
109. A. R. Anderson and E. J. Hart, Radiation Chemistry of Water with Pulsed High Intensity Electron Beams. *J. Phys. Chem.* **66**, 70 (1962).
110. J. K. Thomas, The Nature of the Reducing Species in the Radiolysis of Acidic Aqueous Solutions at High Intensities. *Int. J. Appl. Radiat. Isotopes* **16**, 451 (1965).
111. C. Willis, A. W. Boyd, A. E. Rothwell, and O. A. Miller, Experimental and Calculated Yields in the Radiolysis of Water at Very High Dose Rates. *Int. J. Radiat. Phys.* Chem. **1**, 373 (1969).
112. A. K. Pikaev and P. Ya. Glazunov, Radioliticheskoe vosstannovlennie chetirekh valentnogo ceria pri moshchnostiakh dozy do 10^{23} eV/ml. sek. *Izv. Akad. Nauk SSSR Otd. Khim. Nauk* 940 (1960).
113. A. K. Pikaev, P. Ya. Glazunov, and V. I. Spitsyn, Priblizhennye znacheniia konstant skorosti radiatsionnykh reaktsii s uchastiem gidratirovannogo elektrona. *Dokl. Akad Nauk SSSR* **151**, 1387 (1963).
114. A. K. Pikaev, G. K. Sibirskaia, G. G. Riabchikova, and P. Ya. Glazunov, Mekhanizm obrazovaniia perekisi vodoroda v 0,4 M vodnom rastvore sernoi kisloty pri vysokikh moshchnostiakh pogloshchennoi dozy. *Kine. i Kata.* **6**, 41 (1965).
115. C. J. Hochanadel and J. A. Ghormley, Effect of Temperature on the Decomposition of Water by Gamma Rays. *Radiat. Res.* **16**, 653 (1962).
116. M. Anbar and E. J. Hart, The Activation Energy of Hydrated Electron Reactions. *J. Phys. Chem.* **71**, 3700 (1967).
117. B. Čerček and M. Ebert, Activation Energies for Reactions of the Hydrated Electron. *J. Phys. Chem.* **72**, 766 (1968).
118. B. Čerček, Activation Energy for the Mobility of the Hydrated Electron. *J. Phys. Chem.* **72**, 2279 (1968).

119. B. Čerček, Activation Energies for Reactions of the Hydrated Electron. *Nature* **223**, 491 (1969).
120. K. J. Laidler, "Chemical Kinetics," p. 231. McGraw-Hill, New York, 1965.
121. R. R. Hentz, Farhataziz, D. J. Milner, and M. Burton, γ-Radiolysis of Liquids at High Pressures. I. Aqueous Solutions of Ferrous Sulfate. *J. Chem. Phys.* **46**, 2995 (1967).
122. R. R. Hentz, Farhataziz, D. J. Milner, and M. Burton, γ-Radiolysis of Liquids at High Pressures. II. Aqueous Solutions of Ceric Sulfate. *J. Chem. Phys.* **46**, 4154 (1967).
123. R. R. Hentz, Farhataziz, and D. J. Milner, γ-Radiolysis of Liquids at High Pressures. IV. Primary Yields in Neutral Aqueous Solutions. *J. Chem. Phys.* **47**, 4865 (1967).
124. R. R. Hentz, Farhataziz, D. J. Milner, and M. Burton, γ-Radiolysis of Liquids at High Pressures. III. Aqueous Solutions of Sodium Bicarbonate, *J. Chem. Phys.* **47**, 374 (1967).
125. R. R. Hentz, Farhataziz, and D. J. Milner, γ-Radiolysis of Liquids at High Pressures. V. Reaction of the Hydrated Electron with Water. *J. Chem. Phys.* **47**, 5381 (1967).
126. R. R. Hentz and C. G. Johnson, Jr., γ-Radiolysis of Liquids at High Pressures. VII. Oxidation of Iodide Ion by Hydrogen Atoms in Aqueous Solutions. *J. Chem. Phys.* **51**, 1236 (1969).
127. R. R. Hentz and R. J. Knight, γ-Radiolysis of Liquids at High Pressures. VIII. Primary Yields at 8.7 kbar and Reactions of the Hydrated Electron with H_2O and H_3O^+. *J. Chem. Phys.* **52**, 2456 (1970).

At the onset of a classical chemical reaction, active species are homogeneously distributed in the reaction volume. This is not the case for a chemical reaction induced by radiation. As we have seen in Chapter Two, where we considered the spatial distributions of primary events, primary active species are produced along the path of an ionizing ray only. Classical chemical kinetics, which takes into account changes in concentration of reacting substances as a function of time, is, in principle, inadequate for a kinetic interpretation of the reactions of active species which are initially distributed inhomogeneously. This is why the diffusion-kinetic model has been proposed for chemical reactions in the early stages of water radiolysis. It takes into account the inhomogeneous distribution which varies with time as the active species diffuse away from their point of origin, reacting with each other or with substances present in solution.

As we shall see later, the model is formulated in terms of a system of nonlinear differential equations. Various numerical techniques using fast digital computers have been devised for their calculation. However, the mathematical aspects of the model are beyond the scope of this text. Its purpose is only to present some aspects of the model which contribute to a better understanding of water radiolysis: the visualization of events which still escape a direct experimental observation, as well as the possibility of correlating the experimental data on the effect of reactivity, or of LET, with the yields of primary products of water radiolysis.

It should be noted that the diffusion-kinetic model is quite often the object of vivid controversy. One reason for this is that some reactions of primary species, which the model takes into account, are disputed experimentally; such is the reaction that is assumed to give rise to primary molecular hydrogen (see Chapter Three). Another reason is the arbitrariness or unpreciseness of parameters used in the calculation, such as the initial spatial distribution of primary species, the number of free radicals initially present at the place of energy deposition, or the diffusion constant of primary species. However, it seems that both objections call for caution in using conclusions drawn from the diffusion-kinetic model, but they do not depreciate its contribution to a better understanding of water radiolysis. After all, it should be recalled that the shortcomings mentioned are in fact mainly those of radiation physics and the radiation chemistry of water in general. In appraising the usefulness of this model, account should be taken of the extent to which it enables one to correlate experimental facts with theoretical assumptions which cannot be checked by direct observation, as well as to quantitatively predict new phenomena to be experimentally verified. The diffusion-kinetic model has proved to be of use in both respects.

Since 1953, when Samuel and Magee [1] published the first work dealing with the diffusion-kinetic model of water radiolysis, about twenty relevant papers have appeared. The reader interested in more details on different mathematical, physical, and chemical aspects of the model is referred to the articles of Kuppermann [2–4] and Schwarz [5].

CHAPTER SIX

Diffusion-Kinetic Model

I. Basic Assumptions

The first of the basic assumptions in formulating the diffusion-kinetic model is that the primary species (H, OH, e_{aq}^-, and H_3O^+) present in water after its interaction with radiation are inhomogeneously distributed in space and are in thermal equilibrium with their neighborhood. The inhomogeneity depends on the LET of the radiation. It has been shown (Chapter Two) that the places of localized energy deposit (spurs), which are several thousand angstroms apart so that they have no effect on each other, predominate in the case of ^{60}Co γ-radiation. Heavy charged particles such as, for example, ^{210}Po α-rays, are usually assumed to transfer most of their energy to the spurs which touch each other or even overlap so that they form a cylindrical track. The distribution of primary species is often assumed to be a Gaussian function of the distance with respect to the center of a spherical spur or the axis of a cylinder. The assumptions that localization of energy is brought about only through spherical spurs, or cylindrical tracks, represent only some simplifying approximations to the real situation (Chapter Two, Section IV). It should be noted, however, that proposed improvements [6] contribute more to the correctness of the physical picture for the spatial distribution of primary species than to a better agreement between experimental and theoretical data, as can be seen from the comparison and analysis made by Mozumder and Magee [7]. The objection that primary species are not in thermal equilibrium with their neighborhood also seems to be of no essential importance, especially as the physicochemical and chemical stages of radiolysis may coincide with each other. Similarly, the presence of certain substances, particularly at high concentrations, may have an effect on the process of energy transfer, but for water and dilute solutions this can be disregarded [2].

The second assumption is that primary species diffuse away from the place of energy deposition (that is, from spurs, tracks, etc.) according to Fick's macroscopic diffusion law, $\partial C_i/\partial t = D_i \nabla^2 C_i$. Here, C_i is the concentration of the species i at time t, D_i is the diffusion constant, and ∇^2 is the Laplacian. This assumption seems to be indisputable, since it has been successfully used for other microscopic processes. A major inconvenience is that the diffusion constants of primary species are not reliably established; for the present, only that of e_{aq}^- has been experimentally determined (Chapter Three).

The third assumption concerns the concept of concentration in these considerations and the choice of parameters for the initial distribution of primary species. Namely, in considering the Gaussian distribution of primary species at the site of highly localized energy we have seen that the classical concept of concentration makes no sense here. Hence, the concentration in the diffusion-kinetic model means $c(r, t)$, the probability density of finding a particle at a position in the system determined by vector r at time t after the onset of the chemical stage. For solving the partial nonlinear equations it imports what values are taken for r_0 and N_0, that is, for the radius of initial distribution and for the number of initially present primary species. Direct measurements of these values seem to be unfeasible as yet, and assumptions are made based on experimental data obtained with cloud chambers. Data on the distribution of ions in the tracks of α rays and fast electrons in air were taken from such experiments, and estimations were made for liquid water by means of the Bethe equation (Chapter Two).

The fourth assumption is that the rate laws obeyed by reactions in spurs and tracks are similar to those that would be obeyed if all the reactive species were homogeneously distributed, except that the rate constants for reactions among primary species are permitted to be time-dependent.

The fifth assumption concerns dose rates, that is, energy deposition rates in the case considered. These are supposed to be sufficiently low for the chemical action of radiation to be considered as the sum of actions of individual particles or quanta. This is easily achieved in practice; exceptions are rather rare cases of accelerator irradiations where the dose rates might be so high that overlapping of spurs and tracks occurs (Chapter Five, Section VII).

II. Some General Cases of the Diffusion-Kinetic Model

The rate of change of the probability density $(\partial C_i(r, t)/\partial t)$ at a given point is the algebraic sum of effects which depend on the rate of diffusion of active species toward this point as well as on their production or disappearance rates in reactions with each other or with the solutes. The complexity of a partial differential equation depends on what is taken into account in considering the active species involved—only its disappearance by recombination, or other reactions with solutes and other primary or secondary reactive species. We shall present some general cases, leaving it to the reader to find more details in the original literature.

A. One-Radical

The simplest diffusion-kinetic model is formulated for the case where only the reaction of recombination of free radicals,

$$R + R \longrightarrow RR \quad (1)$$

is taken into account, as was done by Samuel and Magee [1] in the first work published on this subject. This gives the general expression

$$dC/dt = D \nabla^2 C - 2k_1 C^2 \quad (2)$$

where, $C = [R]$, $2k_1$ is the rate constant of the reaction of recombination [Eq. (1)], D is the diffusion constant of the free radical, and ∇^2 is the Laplacian. As can be seen, all other reactions of the free radical are disregarded. The authors have found a solution to the equation by an approximate analytical method, assuming not only that the initial spatial distribution of primary species in the spur is Gaussian but also that it remains such at all times during the expansion of the spur and regardless of their reactions. This was called the prescribed diffusion hypothesis and was subsequently used in calculations by other authors. In spite of drastic simplifications, the model was able to correlate theoretical considerations with the experimental information available at that time on the ratio $(G_{H_2} + G_{H_2O_2})/G_{-H_2O}$ for radiations of very different LET, such as γ-radiation, α particles, and the tritium β-radiation.

B. One-Radical and One-Solute

In considering the effect of scavenger concentration on measured primary yields of molecular hydrogen (Chapter Three) or hydrogen peroxide (Chapter Four) we have assumed that the observed decrease in primary yields with increasing concentration of solute S is due to the competition of the reaction of recombination,

$$R + R \longrightarrow RR \quad [1]$$

with the reaction of the radical with solute S,

$$R + S \longrightarrow RS \qquad (3)$$

If it is supposed that [S] ≫ [R] and, consequently, that changes in concentration of the solute and its diffusion are negligible, then the following expression is obtained for a model which takes into account one radical and one solute:

$$dC/dt = D \nabla^2 C - 2k_1 C^2 - k_3 C_S C \qquad (4)$$

where k_3 is the rate constant of the reaction of the solute with the radical [Eq. (3)], and C_S is the concentration of the solute.

Fricke [8] was the first to propose a solution to the above equation by an approximate analytical method, without comparing the data obtained by this calculation with some known experimental values for water. Schwarz [9] proposed a simplified treatment for solving the equation and compared the theoretical curve with a number of experimental data from his work and from works of other authors. Ganguly and Magee [10] have also given an analytical solution to the above expression and drawn a number of conclusions on the dependence of the recombination reaction of radicals on solute concentration for different high-energy particles.

Flanders and Fricke [11] were the first to use a computer to solve the nonlinear differential equation representing the case where one radical and one solute are involved. They carried out the calculation both for spherical spur and cylindrical track. Some of their conclusions were subsequently used for calculating the corrections for the contribution of the intraspur reaction [Eq. (3)] to the measured stable product yields [12–15].

In considering the reaction of radical with solute, the authors mentioned above assumed that the concentration of solute and its spatial distribution are invariable. It is obvious that this is a simplification, since at the center of spur, or along the axis of cylindrical track, solute concentration is substantially reduced due to high concentration of free radicals. That is why, strictly speaking, the change in solute concentration should also be taken into account as was done by Fricke and Phillips [16]:

$$dC_S/dt = D_S \nabla^2 C_S - k_3 C_S C \qquad (5)$$

where D_S is the diffusion constant of the solute. Their calculation shows that the disregard of the solute concentration change and of the diffusion of solute introduces negligible errors in formulating the spherical spur model, whereas these may amount to 10% and more in the case of cylindrical track [17]. By a somewhat different approach, Kuppermann and Belford [18] also arrived at similar conclusions.

C. Two-Radical and One-Solute

If two radicals are taken into consideration which not only recombine,

$$R_1 + R_1 \longrightarrow R_1R_1 \tag{6}$$

$$R_2 + R_2 \longrightarrow R_2R_2 \tag{7}$$

but also react with each other,

$$R_1 + R_2 \longrightarrow R_1R_2 \tag{8}$$

and with solute,

$$R_1 + S \longrightarrow R_1S \tag{9}$$

$$R_2 + S \longrightarrow R_2S \tag{10}$$

then the real situation in irradiated water is better approximated, but the mathematical formulation becomes more complex. Using the notation defined earlier in this chapter, we obtain the following expressions for reaction rates of R_1, R_2, and S:

$$dC_1/dt = D_1 \nabla^2 C_1 - 2k_6 C_1^2 - k_8 C_1 C_2 - k_9 C_S C_1 \tag{11}$$

$$dC_2/dt = D_2 \nabla^2 C_2 - 2k_7 C_2^2 - k_8 C_1 C_2 - k_{10} C_S C_2 \tag{12}$$

$$dC_S/dt = D_S \nabla^2 C_S - k_9 C_S C_1 - k_{10} C_S C_2 \tag{13}$$

Dyne and Kennedy [19,20] have treated in detail such a complex system. The authors have calculated the yields of primary molecular products and the yield of recombination of radicals into water, which could not be done with the one-radical model. Also very interesting are their observations of variations of the studied yields with changes in the following parameters: reaction rate constants, diffusion constants, and the size of spur. Models of spherical spur and cylindrical track were separately treated, and these considerations gave quite a complete idea of what the diffusion-kinetic model offers, including its arbitrariness and shortcomings.

D. Arbitrariness in Choice of Parameters Used in Solving Diffusion-Kinetic Equations

In considering general cases of the diffusion-kinetic model we saw that for solving diffusion-kinetic equations it is necessary to have values for the rate constants of the reactions studied (k), the diffusion constants of the species observed (D), the number of initially present primary species (N_0), and their initial spatial distribution (r_0). It should be recalled that the first two parameters are directly measurable and that uncertainty in their choice decreases as the amount of new information increases, whereas the other two parameters are still beyond experimental evidence and represent the weak points of

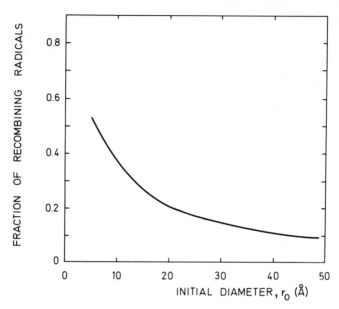

Fig. 6.1. Dependence of the fraction of the initial number of radicals which recombine in a spherical spur on the initial diameter of the spur. (After Kuppermann and Belford [18].)

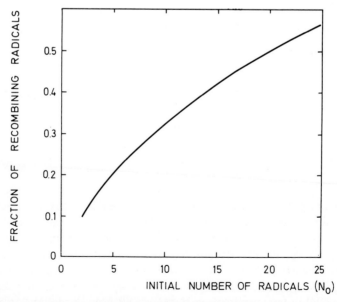

Fig. 6.2. Dependence of the fraction of the initial number of radicals which recombine in a spherical spur on the number of free radicals initially present. (After Kuppermann and Belford [18].)

diffusion-kinetic calculations. It is, therefore, interesting to see to what extent the arbitrariness of the values chosen for r_0 or N_0 may affect the results of calculation and the conclusions drawn from them. The fraction of the initial number of radicals which recombine (also called the fractional molecular yield) in a place of localized energy deposit (such as a spur) is quite convenient for such an observation. It is certain that it should increase with the number of free radicals initially present (N_0) and decrease with increasing initial diameter (r_0). Kuppermann and Belford [18] have studied these variations, and Fig. 6.1 shows the effect of variations of r_0 on the calculated fractional molecular yield. In this calculation it is assumed that 12 radicals are initially present and that their distribution is Gaussian. The values of other parameters used in the calculation were the following: $D_R = 4 \times 10^{-5}$ cm^2 sec^{-1}, $D_S = 4 \times 10^{-6}$ cm^2 sec^{-1}, $k_{R+R} = k_{R+S} = 6 \times 10^9$ M^{-1} sec^{-1}, and $C = 10^{-3}$ M. Figure 6.2 shows the dependence of the same parameter on N_0, the number of radicals initially present in a spur whose r_0 is equal to 10 Å. Other assumptions are the same as those made in calculating data in Fig. 6.1.

III. Theoretical Predictions and Experimental Observations

Experimental information on the primary yield dependence on various parameters (Chapter Five) affords experimental supports for the assumptions made in theoretical considerations in this chapter. The increase in LET of radiation seems really to favor the recombination reactions of primary species and, consequently, as the LET increases the yields of primary molecular products increase with simultaneous decrease in primary radical yields. Also, it was experimentally found that the increase in reactivity of scavenger for a free radical leads to a decrease in the corresponding primary molecular yield. Nevertheless, we also saw that the quantitative correlations of experimental and theoretical data are scarce. Ganguly and Magee [10] treated theoretically the influence of scavenger concentration in more detail and gave the incentive to a systematic experimental work. Schwarz et al. [21] have irradiated systems with particles accelerated by a cyclotron—deuterons of an energy of 18 MeV and helions of energies of 32 and 11 MeV. The systems were adjusted in such a way that measured yields of stable products of radiolysis and usual kinetic treatment enabled G_{H_2} and $G_{H_2O_2}$ to be calculated for different LET values. For the calculation of theoretical curves, the authors made use of a somewhat modified model of Ganguly and Magee, and have obtained a good agreement between calculated and measured values. Burton and Kurien [22] compared the calculated and experimentally obtained values of $G(H_2O_2)$ in aqueous solutions of KCl and KBr subjected to X-radiation of 24 MeV and 50 keV, ^{60}Co γ-radiation, and 3.4-MeV α particles. Here, beside the agreement between experimental and theoretical values, departures calling for a further refinement of the model are pointed out. Also proceeding from the Ganguly–Magee treatment, Burch [23] gave a theoretical analysis and interpretation of the effect of LET on the yields of oxidation of ferrous ions in the Fricke dosimeter and on reduction of ceric ions in the ceric sulfate dosimeter. This analysis is meant to be an illustration of an approach rather than an accurate solution to the problem of chemical yields.

The agreement achieved in the reported cases between calculated and experimentally determined effects was good, although many of the parameters assumed in the above calculations do not appear to be acceptable today, when a greater amount of relevant information is available. This should not surprise us; the reason lies mainly in the simplification of the model and in a considerable number of adjustable parameters, so that almost any experimental results could be fitted. That is why the attempts to treat a many-radical model are of special interest, and the attempts to use the same parameters for interpreting a wide range of very different phenomena (such as the effects of pH, LET, or isotope separation factor) are of particular importance.

A. The Schwarz Many-Radical Model

The procedure used in this calculation [5] is an extension of the Ganguly–Magee treatment to a many-radical system. It is also a modification of the prescribed diffusion approximation, a less time-consuming procedure that gives results within 1% of the exact numerical integration of the equations discussed in Section II.

In the calculation, one proceeds from the assumption that the situation in water in 10^{-11} to 10^{-10} sec from the moment of passage of radiation can be presented in the following way:

$$H_2O \;\;\text{-\!\!\!\sim\!\!\!\sim\!\!\!\sim}\;\; e_{aq}^-, \; H, \; H_3O^+, \; OH, \; H_2 \qquad (14)$$

These species react with each other in the following intraspur reactions, which are known from different studies:

$$e_{aq}^- + e_{aq}^- + 2H_2O \longrightarrow H_2 + 2OH^- \qquad k_{15} = 0.55 \times 10^{10} \; M^{-1} \, \text{sec}^{-1} \qquad (15)$$

$$e_{aq}^- + H + H_2O \longrightarrow H_2 + OH^- \qquad k_{16} = 2.5 \times 10^{10} \; M^{-1} \, \text{sec}^{-1} \qquad (16)$$

$$e_{aq}^- + H^+ \longrightarrow H \qquad k_{17} = 1.7 \times 10^{10} \; M^{-1} \, \text{sec}^{-1} \qquad (17)$$

$$e_{aq}^- + OH \longrightarrow OH^- \qquad k_{18} = 2.5 \times 10^{10} \; M^{-1} \, \text{sec}^{-1} \qquad (18)$$

$$e_{aq}^- + H_2O_2 \longrightarrow OH^- + OH \qquad k_{19} = 1.3 \times 10^{10} \; M^{-1} \, \text{sec}^{-1} \qquad (19)$$

$$H + H \longrightarrow H_2 \qquad k_{20} = 1 \times 10^{10} \; M^{-1} \, \text{sec}^{-1} \qquad (20)$$

$$H + OH \longrightarrow H_2O \qquad k_{21} = 2 \times 10^{10} \; M^{-1} \, \text{sec}^{-1} \qquad (21)$$

III. THEORETICAL PREDICTIONS AND EXPERIMENTAL OBSERVATIONS 179

$$H + H_2O_2 \longrightarrow H_2O + OH \quad k_{22} = 0.01 \times 10^{10} \quad M^{-1} \text{ sec}^{-1} \quad (22)$$

$$H^+ + OH^- \longrightarrow H_2O \quad k_{23} = 10 \times 10^{10} \quad M^{-1} \text{ sec}^{-1} \quad (23)$$

$$OH + OH \longrightarrow H_2O_2 \quad k_{24} = 0.6 \times 10^{10} \quad M^{-1} \text{ sec}^{-1} \quad (24)$$

We see that seven nonlinear differential equations should be formulated to define the behavior of the seven species: the five given by Eq. (14) and two for OH^- and H_2O_2 formed in reactions (19) and (24). Each of these equations describes the behavior of an observed species and reactions in which it is produced or disappears. In view of the increased number of primary species and the greater complexity of the reactions, these equations are also more complex than in the cases considered in the preceding sections. Table 6.1

TABLE 6.1

PARAMETERS FOR INITIAL CONDITIONS USED IN SCHWARZ'S CALCULATIONS[a]

Species	G^0 (yield)	r^0 (Å)	$D, \times 10^{-5}$ cm^2 sec^{-1}
e_{aq}^-	4.78	23	4.5
H	0.62	7.5	7
H_3O^+	4.78	7.5	9
OH	5.70	7.5	2.8
OH^-	0	0	5
H_2	0.15		
H_2O_2	0		2.2

[a] See Schwarz [5].

summarizes the values of different parameters consequentially used in solving these equations as well as those for reactions with solutes. Namely, in addition to the ten reactions listed above, reactions between free radical and solute, $R + S = P$, and corresponding experimental k_{R+S} values are taken into account. In view of what was said in Section II.D about arbitrariness in choice of parameters, it is of interest to dwell on the procedure by which these parameters were chosen.

The diffusion constants of e_{aq}^-, H_3O^+, and OH^- are known from experiments reported in the literature. The values for OH and H_2O_2 were estimated from the self-diffusion constant of water, while that for the H atom was estimated from the diffusion constant of helium.

Some of the G^0 and r^0 values were inferred from a comparison of experimental and calculated values of radiation-chemical yields in simple and well-defined cases. Thus, $G_{H_2}^0$ and $r_{e_{aq}^-}^0$ were found to be 0.15 and 23 Å, respectively, by comparing the calculated and measured values for the effect of nitrite ion on the molecular hydrogen yield, while G_H^0 was determined to be 0.62 by comparing the effects of hydroxide ion and HPO_4^- on the hydrogen atom yield. The radius of the hydroxyl radical spur was found by comparing the calculated and observed effects of nitrite ion on the molecular hydrogen peroxide yield, leading to $r_{OH}^0 = 7.5$ Å. Further, it was assumed that

$$r_{OH}^0 = r_{H_3O^+}^0 = r_H^0 = 7.5 \text{ Å}$$

and $G_{e_{aq}^-}^0$ was determined to be 4.78 by requiring the calculated total net final water decomposition yield $(2G_{H_2} + G_H + G_{e_{aq}^-})$ to be equal to the yield measured at low solute concentration. For the calculation of other G^0 values use was made of the following balancing equations:

$$G_{H_3O^+}^0 = G_{e_{aq}^-}^0$$

$$G_{OH}^0 = G_{e_{aq}^-}^0 + G_H^0 + 2G_{H_2}^0$$

Most of the reaction rate constants are known and taken from the literature. Nevertheless, attention should be drawn to some assumptions which must have been made in this connection. The rate constant for H + OH was estimated by comparison with reactions (20) and (24), since only an indirectly determined value for it was available. Both reactants in reactions (15), (17), and (23) are ions, and this raises problems of ionic strength effects. The rate constant for reaction (15) was determined at 0.2 M ionic strength, which is approximately the initial ionic strength in the spur. For the sake of consistency, all the rate constants were estimated for this ionic strength by introducing a correction by a factor of 0.75. The rate constants for different reactions of the primary species with solutes were obtained by finding the best fits among various reports on relative rate constants and normalizing to absolute values where available. A correction was also introduced in connection with the time-dependence of the rate constants. For fast reactions in liquids proceeding in a time shorter than 10^{-9} sec and at a scavenger concentration of about 1 M the rate constants must be corrected. This effect is due to the finite time required to establish the diffusion-controlled rate.

The number of radicals in spur was taken from the calculation of Mozumder and Magee [24] on the distribution of absorbed energy as a function of the size of spur. The yields were calculated for each type of spur, and the average yield was obtained by weighting the yields by the fraction of energy dissipated in producing each size of spur. According to this calculation, the distribution of absorbed energy as a function of spur size is the following: 56% in the spurs with 1–2 radical pairs, 18.2% for 3–6 radical pairs, 12.8% in short tracks, and the remaining energy in the spurs containing seven and more radical pairs.

The model worked out in this way has been applied to some of those systems which have been claimed to be in disagreement with the diffusion-radical model. As we have seen in Chapter Three, these problems were mainly associated with the origin of atomic and molecular hydrogen (Sections IV and V). Calculations made by using the parameters given in Table 6.1, and corresponding experimental facts, predict various effects consistent with published results. Such effects were the pH dependence of the isotope separation factor for hydrogen production in HDO solutions, the effect of added acid on the reduction of the G_{H_2} with increasing concentrations of nitrate ion or hydrogen peroxide, as well as the $G(N_2)$ increase with increasing nitrous oxide concentrations in solutions containing nitrite. The calculation shows that in most of these cases the effects were too small to be observable under the experimental conditions. In this connection the calculation concerning the time-dependence of the hydrated electron is particularly interesting. It directly concerns the possibility of observing the e_{aq}^- in the spur. Experiments were reported by Thomas and Bensasson [25] in which 3.5-nsec pulses of 3-MeV electrons were used for water irradiations. The change in concentration of solvated electrons was followed by means of fast-absorption spectroscopy; the response time of the system was less than 5 nsec. The light absorption was found to be proportional to the concentration of hydrated electrons. The decay observed was fast at the beginning, with a lifetime of about 70 nsec, followed by a slower decay. Addition of sodium hydroxide (10^{-2} M), which removes H_3O^+, eliminates about 50% of the fast decay, while addition of ethyl alcohol (1 M) removes OH radicals and practically eliminates the remaining 50%. These results point out that the reaction of electrons in spur proceeds slowly over times involving tens of nanoseconds. In the opinion of the authors this was in sharp contrast to the time picture of the diffusion radical model. However, the prediction of the diffusion model shows that for the observation of the time history of the spur by absorption spectroscopy a time resolution better than 10^{-10} sec is required. The same calculation also shows a good agreement between the measured and calculated changes of hydrated electron concentration on a nanosecond scale. It is important to note that in this calculation, as in all other cases, the parameters given in Table 6.1 were used without introducing any new values.

B. Kuppermann's Many-Radical Model

In these calculations [4] the situation in 10^{-11}–10^{-10} sec after the passage of radiation is represented by

$$H_2O \leadsto e_{aq}^-, H, H_3O^+, OH, O \qquad (14a)$$

It can be seen that H_2 is not taken into account here. The oxygen atoms are introduced, although the place they occupy in the scheme is modest. They do not essentially change the general picture; their number was assumed to be less than 4% of the initial number of OH radicals (Table 6.2). The following reactions are taken into account:

Reaction	Rate constant	Ref.
$OH + OH \longrightarrow H_2O_2$	$k_{24} = 0.5 \times 10^{10}\ M^{-1}\ sec^{-1}$	[24]
$OH + e_{aq}^- \longrightarrow OH_{aq}^-$	$k_{18} = 3 \times 10^{10}\ M^{-1}\ sec^{-1}$	[18]
$e_{aq}^- + e_{aq}^- \longrightarrow H_2 + 2OH_{aq}^-$	$k_{15} = 0.5 \times 10^{10}\ M^{-1}\ sec^{-1}$	[15]
$e_{aq}^- + H_3O_{aq}^+ \longrightarrow H + H_2O$	$k_{17} = 2.3 \times 10^{10}\ M^{-1}\ sec^{-1}$	[17]
$OH_{aq}^- + H_3O_{aq}^+ \longrightarrow 2H_2O$	$k_{23} = 14.3 \times 10^{10}\ M^{-1}\ sec^{-1}$	[23]
$H + OH \longrightarrow H_2O$	$k_{21} = 3.2 \times 10^{10}\ M^{-1}\ sec^{-1}$	[21]
$H + e_{aq}^- \longrightarrow H_2 + OH_{aq}^-$	$k_{16} = 3 \times 10^{10}\ M^{-1}\ sec^{-1}$	[16]
$H + H \longrightarrow H_2$	$k_{20} = 1.3 \times 10^{10}\ M^{-1}\ sec^{-1}$	[20]
$e_{aq}^- + H_2O_2 \longrightarrow OH_{aq}^- + OH$	$k_{19} = 1.23 \times 10^{10}\ M^{-1}\ sec^{-1}$	[19]
$H + H_2O_2 \longrightarrow OH + H_2O$	$k_{22} = 0.016 \times 10^{10}\ M^{-1}\ sec^{-1}$	[22]
$OH + S \longrightarrow P$	$k_{25} = 0.6 \times 10^{10}\ M^{-1}\ sec^{-1}$	(25)
$e_{aq}^- + S' \longrightarrow P'$	$k_{26} = 0.6 \times 10^{10}\ M^{-1}\ sec^{-1}$	(26)
$H + S'' \longrightarrow P''$	$k_{27} = 0.6 \times 10^{10}\ M^{-1}\ sec^{-1}$	(27)
$O + H_2O_2 \longrightarrow O_2 + H_2O$	$k_{28} = 0.6 \times 10^{10}\ M^{-1}\ sec^{-1}$	(28)
$O + H_2O_2 \longrightarrow OH + HO_2$	$k_{29} = 0.6 \times 10^{10}\ M^{-1}\ sec^{-1}$	(29)
$O + e_{aq}^- \longrightarrow OH + OH_{aq}^-$	$k_{30} = 0.6 \times 10^{10}\ M^{-1}\ sec^{-1}$	(30)
$O + OH_{aq}^- \longrightarrow HO_{2aq}^-$	$k_{31} = 0.6 \times 10^{10}\ M^{-1}\ sec^{-1}$	(31)
$O + H \longrightarrow OH$	$k_{32} = 0.6 \times 10^{10}\ M^{-1}\ sec^{-1}$	(32)
$O + OH \longrightarrow HO_2$	$k_{33} = 0.6 \times 10^{10}\ M^{-1}\ sec^{-1}$	(33)
$O + S''' \longrightarrow P'''$	$k_{34} = 0.6 \times 10^{10}\ M^{-1}\ sec^{-1}$	(34)

It should be pointed out that the rate constants taken here for the reactions of primary species differ to some extent from those used by Schwarz [5]. There are certain differences in other parameters used in the calculation, as shown in Table 6.2. Here also attention should be drawn to the way in which

TABLE 6.2

Parameters for Initial Conditions Used in Kuppermann's Calculatons[a]

Species	N^0 (radicals)	r^0 (Å)	$D,\ \times 10^{-5}\ cm^2\ sec^{-1}$
e_{aq}^-	2.08	18.75	4.5
H	0.18	6.25	8
H_3O^+	2.08	6.25	10
OH	2.10	6.25	2
O	0.08	6.25	2
OH^-			2
H_2O_2			1.4

[a] See Kuppermann [4].

III. THEORETICAL PREDICTIONS AND EXPERIMENTAL OBSERVATIONS

the parameters of the table were chosen. The diffusion coefficients were selected on the basis of *a priori* considerations and were not subsequently changed. The value used for the diffusion coefficient of e_{aq}^- is close to the published experimental value. The high value of 10^{-4} cm^2 sec^{-1} used for H_3O^+ is based on the assumption of a proton jump mechanism for its diffusion. The diffusion coefficients of OH, OH_{aq}^-, and O were set equal to the self-diffusion coefficient of water (2×10^{-5} cm^2 sec^{-1}), and that of H_2O_2 was obtained by dividing this last by $\sqrt{2}$. The diffusion coefficient for H atoms was obtained by multiplying the experimental value for H_2 by the ratio of the molecular diameters of H_2 and H, according to the Stokes–Einstein diffusion equation.

The initial radii (r^0) and number of radicals (N^0) were chosen by noticing in preliminary calculations that the two most important quantities affecting the final yields are $N_{OH}^0/(r_{OH}^0 \times D_{OH})$ and $r_{e_{aq}^-}/r_{OH}^0$. These were chosen so as to optimize agreement between the theoretical and experimental yields for the ^{60}Co γ-radiolysis of neutral solutions as described by Allen [26]. The mechanism of formation of primary species requires that $r_H^0 = r_{H_3O^+} = r_O^0 = r_{OH}^0$. As can be seen, the total number of radicals under initial conditions is $N_{tot}^0 = 6.5$ radicals.

Taking account of all possible intraspur reactions led to a system of seven simultaneous differential equations involving the 20 reaction steps. They were calculated for spherical spur as well as for several cylindrical tracks of different radical densities. Relative yields were calculated as a function of "pure" LET, that is, for isolated spurs and axially homogeneous tracks of different linear radical densities. These pure LET curves were calculated for different radiations by weighting the pure yields according to the spectrum of energy degradation calculated by Burch [23]. The normalization of relative theoretical yields to absolute ones was done by requiring that G_{OH} be 2.22 for ^{60}Co γ-radiolysis [26]. From what we have seen in Chapter Five, this value is considerably lower than that generally accepted at present. This discrepancy is important more for absolute values than for the trend of calculated data, and should often be taken into consideration when using the theoretical predictions from this work.

Figure 6.3 shows the theoretical curve of the dependence of water decomposition yield on the LET of radiation. There is quite a good agreement between the values for G_{-H_2O} derived from the calculated curve and from various experiments with polonium α-rays, tritium β particles, and photons of different initial energies. The same calculation gives similar predictions for the yields of free radicals, while the trend of the yields of molecular products is reverse, as was to be expected; they increase with increasing LET. A comparison of the calculated data with the values of G_R and G_M derived from experiments with various radiations shows a somewhat less good agreement than in the case of G_{-H_2O}.

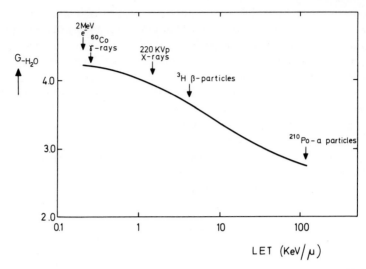

Fig. 6.3. Theoretical curve of the dependence of water decomposition yield (neutral medium) on LET. (After Kuppermann [4]. Reproduced by permission of North-Holland Publishing Company.)

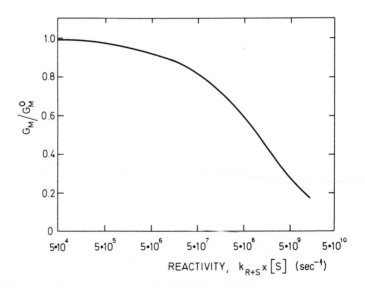

Fig. 6.4. Theoretical curve of the dependence of primary molecular yields on the reactivity; calculated for ^{60}Co γ rays. (After Kuppermann [4]. Reproduced by permission of North-Holland Publishing Company.)

III. THEORETICAL PREDICTIONS AND EXPERIMENTAL OBSERVATIONS

We have seen how the diffusion model assumes that primary molecular products are produced in the recombination reaction of corresponding radicals in spur and that an increase in reactivity causes a decrease in the yield of primary molecular products (G_M) by preventing the recombination reaction. If this drop is presented as a fractional decrease G_M/G_M^0, then a curve should be obtained such as that shown in Fig. 6.4; G_M is the yield measured in the presence of scavenger and G_M^0 is the yield measured in dilute solution, where the presence of scavenger has no effect. The theoretical curve is calculated by using the same parameters (Table 6.2) as those used in calculating the LET effect (Fig. 6.3). The only change is that the rate constant of reaction $e_{aq}^- + S$ was set equal to 0.46×10^{10} M^{-1} sec^{-1}, which is the experimental value for $k_{e_{aq}^- + NO_2^-}$ in neutral solutions [9].

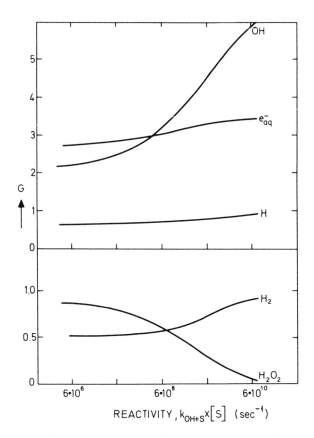

Fig. 6.5. Theoretical predictions on the dependence of primary yields in the γ-radiolysis of neutral water on the reactivity toward OH radicals. (After Kuppermann [4]. Reproduced by permission of North-Holland Publishing Company.)

Data which quantitatively predict that an increase in reactivity toward OH radicals should lead to an increase in the yield of OH, H, e_{aq}^-, and H_2, and not only to a decrease in the yield of H_2O_2, represent an interesting aspect of Kuppermann's calculation [4]. This can be understood by examining the possible intraspur reaction given by Eqs. (15)–(34). We then see that the increased efficiency of reaction (25)—elimination of OH radicals—means not only a depression of reaction (24) in which H_2O_2 is produced, but also a decrease in effectiveness of other reactions of the OH radical, particularly (18) and (21). A consequence of this is that primary reducing species remain in "excess" and thereby their yields, as well as that of the resulting molecular hydrogen, are increased (Fig. 6.5). Similar reasoning holds for an increase in reactivity of scavenger for e_{aq}^-, as can be seen from Fig. 6.6, which is also given

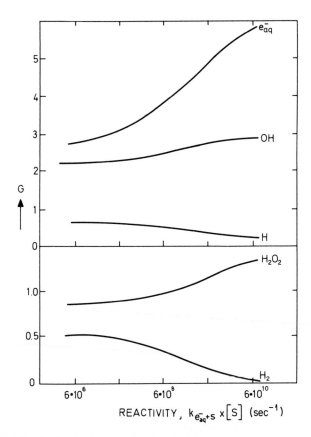

Fig. 6.6. Theoretical predictions on the dependence of primary yields in the γ-radiolysis of neutral water on the reactivity toward e_{aq}^- radicals. (After Kuppermann [4]. Reproduced by permission of North-Holland Publishing Company.)

for ^{60}Co γ-radiation. Such calculations have also been carried out for the polonium α-rays.

The data presented in Figs. 6.5 and 6.6 give incentive to experimental quantiative checking of the model. The absolute values given should be taken with caution, as can be seen if they are compared with those given for primary yields in Chapter Five. However, it is their trend that is crucial for the essence of the model. The plateau values on theoretical curves can be taken as G_R^0 (or G_M^0) values and the ordinate can be represented as the fractional yield (G/G^0). It is surprising that not many attempts of such checks have been reported. We have seen one case in Chapter Four (Fig. 4.4): the increase in reactivity toward e_{aq}^- leads to an increase in $G_{H_2O_2}$ while the increase in reactivity toward OH causes a decrease in $G_{H_2O_2}$ [27]. Curves of best fit to experimental data, obtained with various systems and at different pH values,

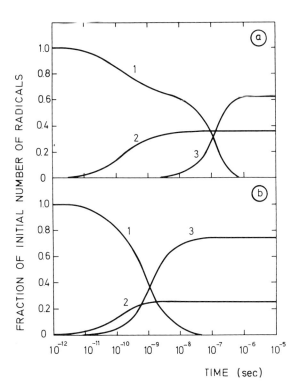

Fig. 6.7. Theoretical diffusion-kinetic curves of the dependence of intraspur events on time. Curves 1, 2, and 3 refer, respectively, to fractions of the initial number of free radicals that survive, those that are built into the molecular product, and those that react with the scavenger. Part a: scavenger concentration is 1×10^{-3} M; Part b: scavenger concentration is 1×10^{-1} M. (After Kuppermann and Belford [18].)

agree fairly well with the theoretical curves presented in Figs. 6.5 and 6.6. It would be interesting to see whether the increase in reactivity toward OH radicals leads to a similar increase of G_{H_2}. Theoretical curves show that these effects should be even more pronounced for high-LET radiations.

Of particular interest for a better understanding of water radiolysis may be the time history of the spur. In Fig. 6.7 the time scale of events occurring in the spur is presented [18]. Radicals are assumed to have a Gaussian distribution. The calculations concern a spherical spur for which the following initial parameters are taken: $N^0 = 12$, $r^0 = 10$ Å, $D_R = 4 \times 10^{-5}$ cm^2 sec^{-1}, and $k_{RR} = k_{RS} = 6 \times 10^9$ M^{-1} sec^{-1}. The fraction of radicals that survive is given by curve 1, the fraction of radicals built into molecular product by curve 2, and the fraction of radicals reacting with scavenger by curve 3. In Fig. 6.7a it can be seen that the reaction of recombination of radicals is for the most part terminated when R + S starts, if the concentration of scavenger is 10^{-3} M. Figure 6.7b shows that a scavenger concentration of 10^{-1} M is sufficient, under the above conditions, for S + R to enter in appreciable competition with R + R. It can also be seen that about 80% of reaction R + R is terminated in a time of about 1 nsec, and that a time on a subnanosecond scale is necessary for observation of the reaction of recombination. A similar conclusion was drawn by Schwarz [5].

References

1. A. H. Samuel and J. L. Magee, Theory of Radiation Chemistry. II. Track Effects in Radiolysis of Water. *J. Chem. Phys.* **21**, 1080 (1953).
2. A. Kuppermann, Diffusion Kinetics in Radiation Chemistry. In "Actions chimiques et biologiques des radiations" (M. Haissinsky, ed.), Vol. 5, p. 85. Masson, Paris, 1961.
3. A. Kuppermann and G. G. Belford, Diffusion Kinetics in Radiation Chemistry. I. Generalized Formulation and Criticism of Diffusion Model. *J. Chem. Phys.* **36**, 1412 (1962).
4. A. Kuppermann, Diffusion Model of the Radiation Chemistry of Aqueous Solutions. In "Radiation Research, 1966" (G. Silini, ed.), p. 212. North-Holland Publ., Amsterdam, 1967.
5. H. A. Schwarz, Application of the Spur Diffusion Model to the Radiation Chemistry of Aqueous Solutions. *J. Phys. Chem.* **73**, 1928 (1969).
6. A. Mozumder and J. L. Magee, Model of Tracks of Ionizing Radiations for Radical Reaction Mechanisms. *Radiat. Res.* **28**, 203 (1966).
7. A. Mozumder and J. L. Magee, A Simplified Approach to Diffusion-Controlled Radical Reactions in the Tracks of Ionizing Radiations. *Radiat. Res.* **28**, 215 (1966).
8. H. Fricke, Track Effect in Iochemistry of Aqueous Solutions. *Ann. Acad. Sci. N. Y.* **59**, 567 (1955).
9. H. A. Schwarz, The Effect of Solutes on the Molecular Yields in the Radiolysis of Aqueous Solutions. *J. Amer. Chem. Soc.* **77**, 4960 (1955).
10. A. K. Ganguly and J. L. Magee, Theory of Radiation Chemistry. III. Radical Reaction Mechanism in the Tracks of Ionizing Radiations. *J. Chem. Phys.* **25**, 129 (1956).

11. D. A. Flanders and H. Fricke, Application of a High-Speed Electronic Computer in Diffusion Kinetics. *J. Chem. Phys.* **28**, 1126 (1958).
12. E. M. Fielden and E. J. Hart, Primary Radical Yields in Pulse-Irradiated Alkaline Aqueous Solution. *Radiat. Res.* **32**, 564 (1967).
13. A. R. Anderson and E. J. Hart, Molecular Product and Free Radical Yields in the Decomposition of Water by Protons, Deuterons, and Helium Ions. *Radiat. Res.* **14**, 689 (1961).
14. E. M. Fielden and E. J. Hart, Primary Radical Yields and Some Rate Constants in Heavy Water. *Radiat. Res.* **33**, 426 (1968).
15. I. G. Draganić, M. T. Nenadović, and Z. D. Draganić, Radiolysis of $HCOOH + O_2$ at pH 1.3–13 and the Yields of Primary Products in γ-Radiolysis of Water. *J. Phys. Chem.* **73**, 2564 (1969).
16. H. Fricke and D. L. Phillips, Application of a High-Speed Electronic Computer in Diffusion Kinetics. II. Depletion Effect in the Spherical One-Radical One-Solute Model. *J. Chem. Phys.* **32**, 1183 (1960).
17. H. Fricke and D. L. Phillips, High-Speed Computations in Diffusion Kinetics. III. Solute Depletion in the Cylindrical One Radical-One Solute Model. *J. Chem. Phys.* **34**, 905 (1961).
18. A. Kuppermann and G. G. Belford, Diffusion Kinetics in Radiation Chemistry. II. One Radical-One Solute Model; Calculations. *J. Chem. Phys.* **36**, 1427 (1962).
19. P. J. Dyne and J. M. Kennedy, The Kinetics of Radical Reactions in the Tracks of Fast Electrons. A Detailed Study of the Samuel-Magee Model for the Radiation Chemistry of Water. *Can. J. Chem.* **36**, 1518 (1958).
20. P. J. Dyne and J. M. Kennedy, A Detailed Study of the Samuel-Magee Model for the Radiation Chemistry of Water. Part II. Kinetics of Radical Reactions in the Tracks of Densely Ionizing Particles. *Can. J. Chem.* **38**, 61 (1964).
21. H. A. Schwarz, J. M. Caffrey, and G. Scholes, Radiolysis of Neutral Water by Cyclotron Deuterons and Helium Ions. *J. Amer. Chem. Soc.* **81**, 1801 (1959).
22. M. Burton and K. C. Kurien, Effects of Solute Concentration in Radiolysis of Water. *J. Phys. Chem.* **63**, 899 (1959).
23. P. R. J. Burch, A Theoretical Interpretation of the Effect of Radiation Quality on Yield in the Ferrous and Ceric Sulfate Dosimeters. *Radiat. Res.* **11**, 481 (1959).
24. A. Mozumder and J. L. Magee, Theory of Radiation Chemistry. VII. Structure and Reactions in Low LET Tracks. *J. Chem. Phys.* **45**, 3332 (1966).
25. J. K. Thomas and R. V. Bensasson, Direct Observation of Regions of High Ion and Radical Concentration in the Radiolysis of Water and Ethanol. *J. Chem. Phys.* **46**, 4147 (1967).
26. A. O. Allen, Radiation Yields and Reactions in Dilute Inorganic Solutions, *Radiat. Res. Suppl.* **4**, 54 (1964).
27. Z. D. Draganić and I. G. Draganić, On the Origin of Primary Hydrogen Peroxide Yield in the γ-Radiolysis of Water. *J. Phys. Chem.* **73**, 2571 (1969).

In considering different radiation-chemical effects we have so far confined ourselves to specifying only the radiation by which they are caused, without dwelling on the radiation source itself and the irradiation technique used. We have seen that by far the largest amount of information on water radiolysis comes from experiments with ^{60}Co γ-radiation and with pulsed electron beams. This chapter gives details on some of these techniques (Sections I and II) as well as on some others which are less frequent in laboratory practice (Sections III and IV). Section V presents details on water purification, cleaning of glass vessels used for irradiation, and some specific aspects of solution preparations.

This chapter deals only with those radiation sources and irradiation techniques which are used at reasearch laboratories; an entire branch of radiation technology, associated with the design of powerful radiation units for technological applications, has been put aside. Those who are interested in this may find useful information in the proceedings of the Conference on Industrial Use of Powerful Sources, which was organized by the International Atomic Energy Agency [1], or in the excellent book edited by Breger [2].

In order that a source of ionizing radiation may be suitable for radiation-chemical research, the following conditions should be fulfilled as far as possible;

1. homogeneous radiation field, ensuring a minimum variation of dose rate in the volume irradiated;
2. well defined and reproducible conditions in the course of irradiation;
3. accurate and reproducible variation of irradiation time, and the possibility of obtaining a wide range of absorbed doses;
4. accurate and reproducible variation of radiation intensity, and the possibility of obtaining a wide range of dose rates;
5. variation of the initial energy of ionizing ray, photon, or particle, and the possibility of obtaining a wide range of LET of radiation.

It is easily understood that a radiation source which would completely satisfy the above conditions is not available; hence, well-equipped laboratories have different radiation sources and use different irradiation techniques. Radiation units with radioactive cobalt make it possible to satisfy conditions 1–3, electron accelerators condition 4, and accelerators of positively charged particles (particularly cyclotrons) condition 5.

CHAPTER SEVEN

Radiation Sources and Irradiation Techniques

I. Radiation Units with ^{60}Co

Radiocative cobalt is obtained by irradiating metallic cobalt with neutrons, through the nuclear reaction $^{59}\text{Co}(n, \gamma)^{60}\text{Co}$. Metallic cobalt is generally irradiated in a nuclear reactor in the form of pellets, small slugs, or thin disks. These are assembled into radiation sources of desired shape and size.

The radioactive decay of ^{60}Co leads to the stable isotope of nickel, $^{60}\text{Co} \xrightarrow{\beta,\gamma} {}^{60}\text{Ni}$, as well as to photons of energies of 1.33 and 1.17 MeV, which are produced in equal numbers. Beta rays, which are also emitted in the radioactive decay of ^{60}Co, practically remain absorbed in the container in which the ^{60}Co is assembled ($E_{\max} = 0.31$ MeV). The half-life of ^{60}Co is 5.3 yr. The radiation intensity decreases by about 1% per month and this should be taken into account in using calibration values for absorbed doses. Table 7.1 gives the correction factors by which a calibration value for the absorbed dose should be multiplied when it is used a month or more after the time at which the calibration was carried out.

A great many radioactive cobalt units are in use at present. Sources ranging from 100 to 10,000 Ci are in common use. The basic problem in constructing the sources is to keep the cobalt adequately shielded while providing means for inserting the sample to be irradiated. This shielding is quite voluminous, as can be seen from data on the thickness of the absorber necessary to reduce the radiation intensity to one half. Half-thickness values (in centimeters) are equal to 11, 5.2, 4.6, and 1.06 for water, concrete, aluminum, and lead, respectively. Lead is usually used for containers in which radioactive cobalt is kept. If such a container has a suitable cavity for inserting and irradiating samples, it is said to be a cavity-type irradiation unit. If

TABLE 7.1

Correction Factors for the Decrease in the Intensity of the Radiation from a ^{60}Co Source

Time, months	Correction factor	Time, months	Correction factor
0	1.000	13	0.867
1	0.989	14	0.858
2	0.978	15	0.849
3	0.967	16	0.840
4	0.957	17	0.830
5	0.947	18	0.821
6	0.936	19	0.813
7	0.926	20	0.804
8	0.916	21	0.795
9	0.906	22	0.786
10	0.896	23	0.778
11	0.887	24	0.769
12	0.877		

radioactive cobalt is taken out of the container for irradiations, and these are carried out in a well-shielded room, then the unit is said to be a cave-type irradiation unit. These are the two kinds of radiation source most frequently encountered; hence, we shall consider them in more detail.

Figure 7.1 shows a cavity-type ^{60}Co source [3]. Six ^{60}Co sources are

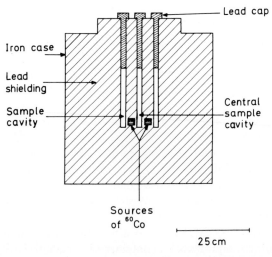

Fig. 7.1. Schematic diagram of a cavity type of ^{60}Co source. (After Draganić *et al.* [*3*].)

I. RADIATION UNITS WITH ^{60}Co

placed in an aluminum carrier. They irradiate one sample placed in the cavity at the central position and six others placed inside the remaining cavities. Samples are introduced through channels which are closed with lead caps. Lead shielding above the cobalt source and the caps serve to stop scattered radiation. Lead bulk is placed in an iron case.

Figure 7.2 shows a cave-type source [4]. It is a circular cave at the middle of which there is a lead container with a multikilocurie ^{60}Co source. This

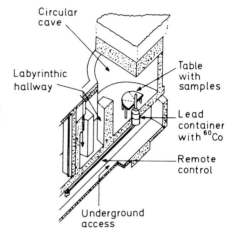

Fig. 7.2. Schematic diagram of a cave type ^{60}Co source. (After Radak and Draganić [4].)

source is pushed by remote control into the operating position at the height of the metal table on which samples to be irradiated are placed. Variations of dose rate of about 10^6 to 10^2 rad hr^{-1} are achieved by placing the sample at different distances from the source. A labyrinthic hallway providing access to the cave reduces radiation intensity, so that a light metal door at the outlet is sufficient. An underground access to the source is provided from the lower, safe side. This is used for lubricating the movable part and, in case of trouble with the remote control, for returning the source from the operating position into the safe position.

One advantage of the cavity-type sources is that they are compact and relatively inexpensive; another is the homogeneity of their radiation field. There are commerical sources where the irradiation cavity is considerably larger (about 1 liter) than that in the source shown in Fig. 7.1. However, the major disadvantage is that the size of samples and equipment that can be put into the cavity is limited. The cave-type source offers a substantially larger room for equipment and makes it possible to irradiate samples at different radiation intensities by varying the distance between the cobalt source and the sample. An inconvenience of this type of source is that a shielding of rather a high cost is needed for its construction.

II. Pulsed Electron Beams

For an accelerator to be suitable for radiation-chemical studies of short-lived species, it is necessary that the radiation pulses produced satisfy some conditions in addition to the general ones mentioned at the beginning of this chapter; they are as follows:

- The pulse length should be negligible as compared to the half-lives of the species studied.
- The rise time and the decay time of the pulse should be short as compared to the pulse duration.
- The energy in the pulse should be sufficient to produce a measurable concentration of the species studied. High energy per pulse and higher concentrations of transients are necessary to study second-order reactions and to minimize first-order reactions and reactions with impurities. Low energy in pulse and low concentrations of intermediates, however, are suitable for studying first-order or slow reactions. In the latter case, high purity of water and high sensitivity of the detection device are also of great importance.

Three types of accelerators are in use at laboratories for producing a pulsed beam; each of these has its advantages and disadvantages [5,6]. The microwave linear accelerator is a traveling wave accelerator. Radiofrequency waves generated by high-power electron tubes (magnetrons and klystrons) are fed to an evacuated, segmented waveguide. Electrons are injected in pulses into the wave guide and accelerated by the electric field of the electromagnetic wave that travels down the tube. Such devices most often give pulses lasting about 1 μsec. A linear accelerator has also been described [7] with a theoretical time resolution of 0.02 nsec. The advantage of the microwave linear accelerators is that they give individual pulses which can be repeated up to several hundred times per second. Moreover, achieving a sufficient electron energy raises no problem, which simplifies the construction of the irradiation vessel as well as the irradiation technique itself and dosimetry. The Van de Graaff accelerator is successfully used for producing pulsed electron beams, although it is better known as a source of light positively charged particles. In principle, electrostatic charge is carried to a high-voltage electrode by means of a rapidly moving belt. The potential difference between the electrode and the ground is used to accelerate electrons to high velocities. The energies of these electrons range mainly from 0.5 to 3 MeV; hence, the irradiation vessel must have quite a thin window in order that the electrons may reach the solution with as little energy loss as possible. Another disadvantage may be the fact that the current of accelerated particles is continuous. To obtain a conveniently pulsed output from the Van de Graaff machine, it is necessary

II. PULSED ELECTRON BEAMS

to have a grid-controlled electron source designed to give a high current. It should be provided with trigger circuits suitable for controlling the pulse duration and repetition frequency. That this does not present any insurmountable difficulty may be seen from the fact that a Van de Graaff generator was used in the first experiments on a nanosecond scale [8]. Technical details may be found in the paper of Ramler *et al.* [9], where a detailed description is given of the electron gun and pulser mounted on the ht terminal of the accelerator producing pulses of 5 A with a duration of 1–100 nsec. Rise and fall times of about 0.3 nsec can be obtained by an interpulse current less than 10^{-11} A. Pulse generators are certainly the simplest and cheapest devices for producing pulsed beams. They consist of pulse-forming networks which are charged in parallel and discharged in series through spark gaps, one or more of which are triggered. In order to make use of the very high surge current which can be obtained from the circuit, the electron source used with this machine is usually based on field emission from a brush of fine wires or some variant of this. A source of this kind can deliver thousands of amperes of electrons with a very short rise time (1 nsec or less). The duration of the pulse is also short (of the order of magnitude of a few nanoseconds), but there are no possibilities of obtaining pulse length variations, in contrast to the case of the Van de Graaff machine. Moreover, this single pulse device does not allow fast pulse repetitions as in the case of the microwave linear accelerator. Furthermore, the energies of the electrons are low (2 MeV or less), which, in view of energy loss in the passage through the wall of the irradiation vessel, means that the range of radiation in aqueous solution is shorter than 7 mm.

Figure 7.3 is a schematic presentation of a setup for radiation-chemical experiments with pulsed electron beams; it consists of an accelerator which is

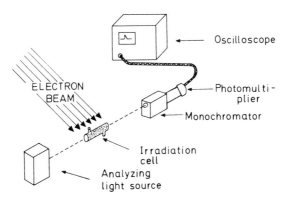

Fig. 7.3. Schematic representation of an arrangement for pulse radiolysis using spectrophotoelectric records.

the source of radiation, an irradiation cell, and a detection device. Details on different setups may be found in papers of Boag [6] and Hart [10] as well as in the books of Matheson and Dorfman [5] and Pikaev [11]. We shall only make some general comments. The irradiation cells and the detection equipment are in many respects similar to those used in flash-photolysis [12–14] and designed for optical studies of short-lived species. The irradiation cell, especially the windows, should be made of glass resistant to radiation-induced coloration, such as high-purity fused silica. The construction of the cell is often such as to ensure the cell's being not displaced when solutions are changed [10], and even to make it possible to change solutions by remote control from outside the irradiation area [15]. Since the irradiation itself is of very short duration, such designs enable the samples studied to be rapidly changed and thereby ensure a high efficiency in producing experimental results. Cell designs for a specific purpose are much more complex. Such is, for example the one for radiolysis at gas pressures as high as 100 atm [16] or the one ensuring long optical paths (up to 80 cm), although the cell itself is only 4 cm long. This is a multiple-reflection cell assembly [17], which makes it possible to work with low concentrations of reactive species or to study those which otherwise could not be observed because of low values of their molar extinction coefficients. The source of light depends in the first instance on the type of experiment. For spectrographic recording, quite powerful xenon-filled spectroflash lamps are used. These ensure an intense continuum with few superposed emission lines. They should also provide a flash of a duration close to that of electron pulse. This is desirable in order to achieve a maximum time resolution of transient spectra. If the characteristic spectrum of the species studied is known, then for study of its kinetic behavior use is made of spectrophotoelectric records. In this case, the change of absorption with time is followed photoelectrically at a selected wavelength, and the light intensity should be as constant as possible over the period of measurement. For this, use can be made of dc tungsten filament lamps, but they are limited mainly to visible and infrared regions. High-pressure mercury arcs give light of very high intensity, so that Čerenkov radiation and other "noises" are not so important. However, for the measurements of entire spectra high-pressure xenon arcs are the most convenient. The source of light can be placed as shown in Fig. 7.3, that is, so that the light beam falls on the vessel perpendicularly to the electron beam, but the irradiation cell is often placed between the electron source from the one side and the light source from the opposite side. Then the light beam falls in parallel to the electron beam. After the light has traversed the solution once, or by means of a suitable mirror arrangement many times, it goes into a monochromator where a given wavelength is selected and is transmitted into a photomultiplier. The signal thus produced, amplified if necessary, is displayed on an oscilloscope and photographed by a polaroid camera. The choice of photodetector and oscilloscope also depends on the system studied. Electron noise problems encountered in experiments on pulse radiolysis are rather serious [15], especially in nanosecond-level experiments [8,9]. Measurements in the uv region are also difficult. Strong Čerenkov radiation may often be very disturbing. A successful attempt in eliminating various inconveniences for measurements at 2000 Å has been reported by Christensen et al. [18].

Bronskill and Hunt [7] have described a new technique for observing transient absorption signals as short-lived as 0.02 nsec. This technique makes use of the Čerenkov radiation produced by a 30 MeV electron beam from a linear accelerator to detect the absorption of transient species produced by the fine-structure pulses (0.01 nsec) of the electron beam. A stroboscopic effect is produced by varying the phase difference between Čerenkov radiation flashes and the fine-structure electron pulses. With this technique a conventional detection system can be used to achieve a time resolution of 2×10^{-11} sec.

Only a few pulse-radiolysis studies with conductivity measurements have been carried out on water and aqueous solutions. A study of the mobility of the hydrated electron also gives a description of an apparatus for such conductivity measurements [19].

III. POSITIVELY CHARGED PARTICLES

As we have seen in the preceding chapter, only a few experiments with positively charged particles have been performed. The low penetrability of these particles in condensed medium raises considerable problems as to experimental technique and dosimetry. The obtaining of results is much more complicated and time-consuming than with the radiation sources described in the preceding two sections. Some radioactive isotopes and different types of accelerators are used as sources of positively charged particles. Facilities offered by the latter, in particular by cyclotrons, are considerable; unfortunately, these expensive machines are seldom at the disposal of chemical laboratories.

Polonium-210 emits α rays of an energy of 5.30 MeV. Their range in water is only 39 μ, and most experiments with these rays are carried out in such a way that a suitable polonium salt is dissolved in the solution studied. Since polonium is extremely poisonous, its solutions should be handled with great care, conformably to the safety measures anticipated for chemical operations with corresponding radioactivities of α emitters. Corrections for radioactive decay are much larger here than in the case of radioactive cobalt, since the half-life of ^{210}Po is 138 days.

Nuclear reactions may serve as much more suitable internal sources of positively charged particles. At present, use is made of some nuclear reactions induced by thermal neutrons, in which the struck nucleus decays into positively charged fragments whose LET is close to or higher than that of α rays in the preceding case [20–24]. These reactions are as follows:

$$^6Li(n, \alpha)^3H, \quad ^{10}B(n, \alpha)^7Li, \quad \text{and} \quad ^{235}U(n, \text{fission fragments})$$

In these cases the total kinetic energy is about 4.8, 2.4, and 185 MeV, respectively. An advantage of these techniques is a simpler handling, since in the first two cases solutions are practically nonradioactive. The low radioactivity of the fission fragments also does not require any special safety measures in experiments with low absorbed doses, as is the case in fundamental studies of water radiolysis. Homogeneous distribution of radiation sources in solution and the possibility of varying dose rates by varying the concentration of target nuclei are certainly considerable advantages of these techniques for studying reactions induced by radiation of high LET. In such experiments solutions are irradiated in quartz ampuls, since other kinds of glass become considerably radioactive under the action of thermal neutrons. Irradiations are carried out in a nuclear reactor; it is desirable to do this in a thermal-neutron irradiation facility which provides highly thermalized neutrons with a very low γ-ray background. For the calculation of the absorbed dose it is necessary to know the neutron flux under the irradiation conditions. Furthermore, account should be taken of the action of the accompanying radiation component with low LET consisting of γ rays and recoil protons, which is inevitable under such irradiation conditions. A work of Sigli and Pucheault [24] may be of great help for those interested in various technical details which should be taken into consideration in using nuclear reactions as internal sources of positively charged particles.

Accelerators are used as external sources. Of the different accelerators designed for nuclear research, the Van de Graaff machine, the cyclotron, the Cockcroft–Walton, and the linear ion accelerators are the most widely used among chemists. The nuclei of light elements are accelerated in the linear ion accelerator by passing through a series of tubes of increasing length to which an alternating potential is applied; the particles acquire an increment of energy each time they traverse the gap between two tubes. Schuler [25] made use of such a device for irradiations with C^{6+} ions whose initial energy was about 120 MeV. The Cockcroft–Walton accelerator is, in principle, a voltage multiplier with a series of rectifiers and condensers that rectify and multiply the output voltage of a transformer. The potential difference produced in such a way is usually less than 1.5 MeV. It was used for irradiations with protons whose initial energy was 1.2 MeV or less [26]. In the cyclotron, light ions are introduced into the center of the gap between two hollow, evacuated semi-

III. POSITIVELY CHARGED PARTICLES

circular accelerating electrodes (dees) placed between the poles of an electromagnet designed to produce a constant field. By rapidly alternating the potential applied to the two dees, the ions are accelerated along a spiral path which increases in diameter as the energy of the ions increases. The accelerated ions, deuterons, or helions, are extracted at the outlet of the dees as a continuous beam.

Figure 7.4 shows schematically a setup for irradiation with charged particles produced in an accelerator [27,28]. A solution is placed in a cell

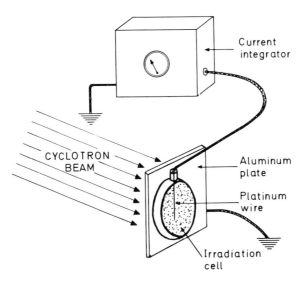

Fig. 7.4. Schematic representation of an arrangement for radiation-chemical studies with positively charged particles from an accelerator.

which, in the case of weakly penetrating particles, such as those from a Van de Graaff or Cockcroft–Walton accelerator, must have a thin window made of glass or mica. The cell is placed against the window of the machine. A platinum wire is immersed in the solution and connected to an appropriate current meter, the output going to ground. The particle beam is entirely stopped in the solution and all the charge must go to ground through the meter. The number of particles that reached the solution is found by integrating the metered current over the time. From this data and the value of the energy of the particles that reached the solution, the dose absorbed is calculated. While the thickness of absorber traversed by the beam before it reaches the solution can give, by a simple calculation, the value of the particle energy, the value for current read on the current meter requires different corrections. One of these is due to the existence of electrical pathways from the solution to ground other than the meter, such as, for example, ionized air between the window of the accelerator and the window of the irradiation cell. Another is due to the fact that the impact of the beam on the window of the irradiation cell gives rise to secondary electrons which, among other things, reduce the positive charge by getting into the solution. Details on the setup for irradiation and measurement of the currents and energies of deuterons (up to 20 MeV) and helions (up to 40 MeV) from a cyclotron may be found in the paper of Schuler and Allen [27]. Hart et al. [29] have given very useful technical data for work with protons (0.3–2 MeV) from a Van de Graaff accelerator and deuterons (3.5–21.2 MeV) from a cyclotron. Pucheault et al. [30] have given a description of a setup for irradiation with protons of low energies (less than 0.6 MeV). It should be noted that many of the arrangements described also take care of mixing the solution during irradiation. This is understandable, since we are dealing with external sources whose radiation range in liquid is most often of the order of magnitude of a millimeter, or substantially less. Under the condition that dose rates are sufficiently low, the mixing should ensure the renewal of the liquid layer in close proximity to the window of the irradiation cell.

IV. Some Other Irradiation Techniques

In some radiation-chemical experiments use was also made of internal β sources. The aqueous solutions studied contained solutes labeled with the radioactive isotopes ^{32}P or ^{35}S [31,32]. Tritium β rays also served for inducing chemical changes in aqueous solutions, where the water was enriched with the heavy isotope ^3H [33–35]. The energies of β-radiations from these three radioactive isotopes conveniently complete the range of LET in the transitional region; the β-radiation from radioactive phosphorus is close in its LET value to the radiation from radioactive cobalt, while the LET value of radioactive sulfur or tritium is larger. Table 7.2 gives some data on β emitters of

TABLE 7.2

Beta Emitters as Internal Sources

Radioactive isotope	Decay			Half-life	β Energy, MeV		Range in H_2O, mm
					Maximum	Average	
^{32}P	^{32}P	$\xrightarrow{\beta}$	^{32}S	14.2 days	1.71	0.70	7.9
^{33}S	^{35}S	$\xrightarrow{\beta}$	^{35}Cl	87.2 days	0.167	0.049	0.32
^3H	^3H	$\xrightarrow{\beta}$	^3He	12.26 yr	0.018	0.0055	0.0055

interest for experimental work. We see that all the three isotopes are pure β emitters and that their radioactive decays lead to stable isotopes. Safety measures for work with these isotopes are much less strict than those for work with polonium sources, since they represent a much smaller health hazard.

Conventional X-ray devices have served as external sources of radiation [36–39]. The weak penetrability of the so-called soft X-radiation also raises the problems mentioned in the preceding section in connection with positively charged particles of low energies. A conventional X-ray device consists of a tungsten wire cathode and a massive anode, also made of tungsten, placed inside an evacuated glass jacket. The cathode is heated to provide a stream of electrons which are accelerated toward the anode by a potential applied to the cathode and anode. X Rays are produced when high-velocity electrons are abruptly decelerated by striking a tungsten target which serves as the anode. The energies of the photons range from values close to zero up to that corresponding to the maximum energy of the incident electrons. Commercial X-ray devices are designed to have maximum operating potentials in the range 40–300 kV_p, where kV_p is the kilovoltage peak of the X-ray power supply. They provide X-ray beams in which most photons have energies ranging from about 10 to 100 keV. It is assumed that most photons in a non-filtered beam have energies numerically close to one third of the maximum operating voltage. These relatively low photon energies lead to the fact that the radiation field varies over even a small volume and that the irradiation technique and dosimetry are not as simple as in the case of the γ-radiation from a ^{60}Co source. Different types of robust X-ray devices designed for irradiations under different conditions have been used for a long time in Soviet laboratories for a great variety of experiments [40].

Intermittent irradiation is not used in the radiation chemistry of water as frequently as in photochemistry. Hummel *et al.* [41] have described a rotating-sector method applied to reactions induced by the γ rays from a 1000 Ci ^{60}Co source. The sector was placed between the source and samples, and connected through a series of pulleys to a variable-speed motor. It was a solid steel cylinder, about 15 cm in diameter and 30 cm long, with two 60° sectors cut out on opposite sides.

Nuclear reactors are rarely used for quantitative studies. The reason for this is the complexity of pile radiation. It consists of (a) fast neutrons from fission that are being slowed down and thermalized by collisions with the atoms of the materials present, (b) γ rays which come chiefly from fission and from fission products, and (c) β rays that are mainly due to the radioactivity induced by neutrons in the sample studied. The spectra of neutron and γ energies vary with different types of reactors as well as from point to point within the same reactor. Usually they are known only incompletely. Since radiation intensity also may vary in the course of irradiation, the problem of dosimetry is of particular importance. It should be recalled that a contribution to the dose absorbed in an aqueous sample is also given by fast neutrons via recoil protons which have different energies and LET values, corresponding to those between the γ and α rays. This also contributes to the fact that the mixed radiation is less convenient from the point of view of dosimetry and that interpretation of results is more difficult than in the case of radiations considered before. However, as mentioned in the preceding section, neutron-induced nuclear reactions may be a very useful tool in studying chemical changes caused by high-LET radiations. The kinetics of ceric ion reduction reported by Sigli and Pucheault [24] show that these measurements can be quantitatively carried out in a mixed pile radiation field as well as with any other source of radiation.

A continuous electron beam from a Van de Graaff generator is a suitable radiation source for irradiations at high absorbed doses over a wide range of dose rates, since current variations allow for dose rate variations by a factor of a million and more. Many technical details that are useful for quantitative work are given in the paper of Saldick and Allen [28]. The microwave linear accelerator also may serve for attaining large doses at very different dose rates. Irradiations are carried out here with constant pulses which are repeated up to one hundred and more times per second; dose rate variations are achieved by varying the distance of the sample irradiated from the beam outlet [42].

Atomic hydrogen can be produced in electrodeless high-frequency discharge. Czapski and Stein [43] have given a description of a device used in their studies of the behavior of the H atom. The discharge tube was made of quartz, since the temperature of the tube during discharge exceeds the mollifying point of Pyrex. The heat released during discharge was removed by cooling with an air blower. Light shields prevented the uv radiation produced in discharge from reaching the reaction vessel. Carefully purified molecular hydrogen was stored in a large reservoir. This maintained a constant pressure of gas which was supplied to the discharge tube through a reducing valve. The frequency chosen for the discharge was 27–30 Hz. The total possible power input was 2000 W dc, with an rf power of 1400 W maximum in the rf discharge. The HF coil was a copper tube through which cooling water was circulated.

A compact apparatus for photogeneration of hydrated electrons has been described by Schmidt and Hart [44]. In principle, the device makes use of the photoreaction $OH^- + h\nu \rightarrow OH + e_{aq}^-$, which occurs in a H_2-saturated solution at pH 11 and is followed by the reactions $OH + H_2 \rightarrow H + H_2O$ and $H + OH^- \rightarrow e_{aq}^-$. Hence, each quantum of uv radiation can produce two hydrated electrons whose reactions with different substances can be followed by observing the change in absorption of light in a solution using a special flash spectrometric array (described in the paper). In a single 40-μsec light pulse, 10^{-7} moles of e_{aq}^- are produced. The sensitivity of the instrument is 10^{-9} moles of hydrated electrons, and it makes possible different studies on the properties of e_{aq}^- under well-defined experimental conditions without using costly accelerators.

V. Preparation of Samples for Irradiation

As we have seen in the preceding chapters, the study of water radiolysis most often concerns the behavior of reactive species at very low concentrations—millimoles and much lower. It is therefore understandable that very high requirements on purity of water, cleanness of irradiation vessels, and care in preparation of solutions are imposed in such a study. Any impurity present competes with reactive species or solutes observed and might influence the results obtained.

Most water purification systems used in standard chemical laboratories aim to attain a low conductivity, ensuring effective removal of inorganic impurities. However, water of a sufficiently low conductivity may still contain some organic impurities and be unsuitable for radiation-chemical studies. This is why water purification for radiation-chemical research is often carried out by triple distillation of ordinary distilled water: first from an acid dichromate solution, then from alkaline permanganate, and finally without any additive into a silica container (Fig. 7.5). This is done in a continuous system in a gas (nitrogen or oxygen) stream. Each distillation is done through a column packed with glass helices. The entire device is made of glass, and the water never comes in contact with rubber or plastics. For further purification the following procedure has been proposed [45]: Samples of air-saturated triply distilled water were irradiated with a ^{60}Co source until a concentration

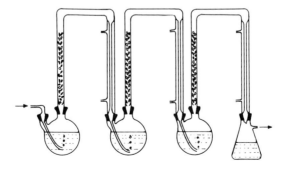

Fig. 7.5. Purification of water for radiation-chemical experiments.

of about 10^{-4} M H_2O_2 was attained. The water was then poured into a carefully cleaned and steamed-out tube of fused silica and photolyzed by placing it inside the coil of a helical low-pressure mercury resonance lamp, until all the peroxide was destroyed. It was found that small concentrations of hydrogen peroxide, about 3–4 μmoles, still remain in the solution. This was ascribed to the action of the 1849-Å light component. A complete destruction of H_2O_2 can be achieved if the sample photolyzed is surrounded by a filter consisting of a solution of KI (0.155 gm liter^{-1}) and I_2 (0.108 gm liter^{-1}), to cut out the far uv. A check of water purity may be carried out by measuring an appropriately chosen radiation-chemical yield. The measurement of the following yields might be convenient:

$$G(H_2) = 0.45 \pm 0.02 \quad \text{and} \quad G(H_2O_2) = 0.85 \pm 0.04$$

These should be measured in deaerated water containing about 10^{-4} M KBr and in air-saturated water containing 1×10^{-4} M KBr, respectively.

It is obvious that such care in purifying water must be accompanied by equal care in cleaning irradiation vessels. These are cleaned by immersion in a boiling nitric acid–sulfuric acid mixture followed by thorough rinsing with triply distilled water. This is not quite sufficient, and one or more of the following procedures are used in addition to the preceding: heating in a furnace at a temperature somewhat lower than that at which the glass begins to mollify; cleaning with a current of steam obtained by boiling purified water; irradiation of the cleaned vessel full of triply distilled water with a dose of about 1 Mrad of ^{60}Co γ-radiation. The purpose of these procedures is to remove trace amounts of grease and organic substances which seem to bury themselves into submicroscopic fissures, where they are not accessible to cleaning with acid mixtures. On γ-irradiation, these materials come into the water and are destroyed by reactions with free radicals. Perfectly cleaned vessels are best kept full of triply distilled water, since it seems that glassware exposed to ordinary air picks up a film of grease or other impurities. Irradiation vessels must not be closed with another material (for example, rubber, cork, or plastics). It is also desirable that ground glass stoppers be female joints and that the tubes end in a male joint.

Degassing of solutions and simultaneous filling of many samples can be carried out by means of a vessel such as that shown in Fig. 7.6. Here the irradiation vessels are Pyrex cylindrical ampuls 60 mm long and 18 mm in diameter. They end in a normal B_5 joint located at the end of a narrow neck 45 mm long and 2 mm in diameter. Six to eight ampuls are placed at the top of a 250-cm^3 flask, which is connected through a cold trap to a standard high-vacuum pump. By alternately shaking and pumping, the air is removed from the solution which fills somewhat more than half the vessel. By simply turning the vessel, the ampuls are filled with degassed solution up to the

V. PREPARATION OF SAMPLES FOR IRRADIATION

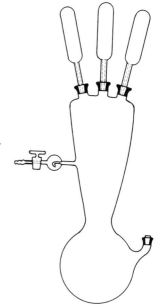

Fig. 7.6. A vessel for solution degassing and simultaneous preparation of many samples.

top, so that no gas phase remains, and then closed with stoppers also full of solution. For a short irradiation, which is most often the case, such a closure of ampuls is sufficient. They are sealed only if the sample must for some reason await irradiation for a long time, or if the irradiation should last a somewhat longer time. Such a device for degassing is also used to introduce a known amount of gas (O_2, N_2O, etc.) into a previously degassed solution.

Syringe technique affords a simple and precise handling of aqueous samples [*10*]. Solutions of known composition can be prepared by injecting the content from one syringe into the other. Dilution can also be carried out by transferring water, degassed if necessary, into an appropriate volume of a concentrated solution which is placed in a larger syringe (50 or 100 ml). Commercial hypodermic syringes are used (Fig. 7.7). The solution is transferred from one syringe (a) to the other through a capillary tube (b). A flat glass disk (c) is placed in the syringe where the mixing is performed; by repeated inversions of the syringe it permits an efficient mixing.

Preparation of alkaline solutions calls for special care: The CO_2 content of hydroxide must be as low as possible, since it reacts very effectively with the primary products of water radiolysis. The method of preparing hydroxide free of CO_2 that is used in standard chemical laboratory practice is not very suitable, since it requires that the solution be prepared long before use,

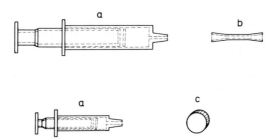

Fig. 7.7. Syringes for the preparation and dilution of solutions.

whereas for study of radiolysis it is better to have freshly prepared solutions. For this purpose [46] analytical-grade metallic sodium is first rinsed with diethyl ether and then quickly cut. Waste parts are removed and pieces of about 0.5 gm are dissolved in triply distilled water. This is done in a nitrogen atmosphere; the flask is kept in an ice-cold bath. After dissolving the quantity needed, the concentration of the hydroxide prepared is determined by using an automatic burette connected to the flask. The burette also serves to introduce the given amount of freshly prepared hydroxide into the vessel shown in Fig. 7.6. The transfer of the solution with adjusted pH is done in a nitrogen atmosphere. Before filling the ampuls, the solution is degassed or a gas is introduced. In certain cases metallic sodium is distilled directly in water, but the procedure is rather time-consuming and seems justified only for concentrated solutions, 1 M NaOH or more.

References

1. "Industrial Uses of Large Radiation Sources," Vol. II. Int. At. Energy Ag., Vienna, 1963.
2. A. H. Breger, ed. "Osnovy radiatsionno-khimicheskogo apparato-stroenia," Atomizdat, Moskva, 1967.
3. I. G. Draganić, B. V. Radosavljević, and N. J. Milašin, A Laboratory ^{60}Co Source of 60 Curies. *Bull. Nucl. Sci. Inst. Boris. Kidrič* (*Belgrade*) **7**, 151 (1957).
4. B. B. Radak and I. G. Draganić, The "2kC" ^{60}Co Radiation Unit of the Boris Kidrič Institute at Vincha. *Bull. Nucl. Sci. Inst. Boris Kidrič* (*Belgrade*) **13**, 77 (1962).
5. M. S. Matheson and L. M. Dorfman, "Pulse Radiolysis." MIT Press, Cambridge, Massachusetts, 1969.
6. J. W. Boag, Physical Methods in Radiation Chemistry and Radiobiology. *In* "Actions chimiques et biologiques des radiations" (M. Haissinsky, ed.), Vol. 6, p. 4. Masson, Paris, 1963.
7. M. J. Bronskill and J. W. Hunt, A Pulse Radiolysis System for the Observation of Short-Lived Transients. *J. Phys. Chem.* **72**, 3762 (1968).
8. J. W. Hunt and J. K. Thomas, Pulse Radiolysis Studies Using Nanosecond Electron Pulses: Observation of Hydrated Electrons. *Radiat. Res.* **32**, 149 (1967).

9. W. J. Ramler, K. Johnson, and T. Klippert, High Current Pulsed Electron Source—Van de Graaff. *Nucl. Instrum. Methods* **47**, 23 (1967).
10. E. J. Hart, The Hydrated Electron. In "Actions chimiques et biologiques des radiations" (M. Haissinsky, ed.), Vol. 10, p.11. Masson, Paris, 1966.
11. A. K. Pikaev, "Impulsnyi radioliz vody i vodnykh rastvorov," Nauka, Moskva, 1965.
12. E. G. Niemann and M. Klenert, A New Flash-Photolysis System for the Investigation of Fast Reactions. *J. Phys. Chem.* **72**, 3766 (1968).
13. J. W. Boag, Techniques of Flash Photolysis. *Photochem. Photobiol.* **8**, 565 (1968).
14. G. Porter and M. R. Topp, Nanosecond Flash Photolysis and the Absorption Spectra of Excited Singlet States. *Nature* **220**, 1228 (1968).
15. J. P. Keene, Pulse Radiolysis Apparatus. *J. Sci. Instrum.* **41**, 493 (1964).
16. M. S. Matheson and J. Rabani, Pulse Radiolysis of Aqueous Hydrogen Solutions. I. Rate Constants for Reaction of e_{aq}^- with Itself and Other Transients. II. The Interconvertibility of e_{aq}^- and H. *J. Phys. Chem.* **69**, 1324 (1965).
17. J. Rabani, W. A. Mulac, and M. S. Matheson, The Pulse Radiolysis of Aqueous Tetranitromethane. I. Rate Constants and the Extinction Coefficient of e_{aq}^-. II. Oxygenated Solutions. *J. Phys. Chem.* **69**, 53 (1965).
18. H. C. Christensen, G. Nilsson, P. Pagsberg, and S. O. Nielsen, Pulse Radiolysis Apparatus for Monitoring at 2000 Å. *Rev. Sci. Instrum.* **40**, 786 (1969).
19. K. H. Schmidt and W. L. Buck, Mobility of the Hydrated Electron. *Science* **151**, 70 (1966).
20. W. R. McDonell and E. J. Hart, Oxidation of Aqueous Ferrous Sulfate Solutions by Charged Particles. *J. Amer. Chem Soc.* **76**, 2121 (1954).
21. L. Ehrenberg and E. Saeland, Chemical Dosimetry of Radiations Giving Different Ion Densities. An Experimental Determination of G Values for Fe^{2+} Oxidation, Jener Publications, No. 8, Kjeller, Norway, 1954.
22. I. Draganić and J. Sutton, Effets chimiques de l'irradiation par des particules lourdes: emploi du système sulfate ferreux–sulfate de lithium comme dosimètre simultané de neutrons thermiques et de radiations ionisantes. *J. Chim. Phys.* **52**, 327 (1955).
23. R. H. Schuler and N. F. Barr, Oxidation of Ferrous Sulfate by Ionizing Radiations from (n, α) Reactions of Boron and Lithium. *J. Amer. Chem. Soc.* **78**, 5756 (1956).
24. P. Sigli and J. Pucheault, Radiolyse de solutions aqueuses par des rayonnements à ionisations denses. I. Radioréduction induite par la réaction nucléaire $^{10}B(n, α)^7Li$ dans des solutions sulfuriques de Ce^{IV} en présense de Tl^I et de Ce^{III}. *J. Chim. Phys.* **65**, 1543 (1968).
25. R. H. Schuler, Radiation Chemical Studies with Heavy Ion Radiations. *J. Phys. Chem.* **71**, 3712 (1967).
26. I. Draganić, M. Simić, N. Milašin, and B. Radak, The Effect of Low Energy (Less than 0.7 MeV) Protons on Aqueous Solutions. I. Oxidation of Ferrous Ions in Fricke Dosimeter, A Portion of *Proc. 1st All-Union Conf. Radiat. Chem., Moscow, 1957*, p. 57. Consultant Bureau, Inc., New York, 1959.
27. R. H. Schuler and A. O. Allen, Absolute Measurement of Cyclotron Beam Current for Radiation-Chemical Studies. *Rev. Sci. Instrum.* **26**, 1128 (1955).
28. J. Saldick and A. O. Allen, The Yield of Oxidation of Ferrous Sulfate in Acid Solution by High-Energy Cathode Rays. *J. Chem. Phys.* **22**, 438 (1954).
29. E. J. Hart, W. J. Ramler, and S. R. Rocklin, Chemical Yields of Ionizing Particles in Aqueous Solutions: Effect of Energy of Protons and Deuterons. *Radiat. Res.* **4**, 378 (1956).
30. J. Pucheault, R. Julien and J. Chevrel, Dispositif de sortie dans l'air et contrôle d'un faisceau de protons de faible énergie ($E < 600$ keV). *Rev. Phys. Appl.* **2**, 137 (1967).

31. T. J. Hardwick, Radiation Chemistry Investigation of Aqueous Solutions Using P^{32} and S^{35} as Internal Sources. *Can. J. Chem.* **30**, 39 (1952).
32. N. Miller and J. Wilkinson, II. Actinometry and Radiolysis of Pure Liquids. Actinometry of Ionizing Radiation. *Discuss. Faraday Soc.* **12**, 50 (1952).
33. T. J. Hardwick, The Effect of the Energy of the Ionizing Electron on the Yield in Irradiated Aqueous Systems. *Discuss. Faraday Soc.* **12**, 203 (1952).
34. E. J. Hart, Molecular Product and Free Radical Yield of Ionizing Radiations in Aqueous Solutions. *Radiat. Res.* **1**, 53 (1954).
35. E. Collinson, F. S. Dainton, and J. Kroh, The Radiation Chemistry of Aqueous Solutions. II. Radical and Molecular Yields for Tritium β-Particles. *Proc. Roy. Soc. (London)* **A265**, 422 (1962).
36. H. Fricke and E. J. Hart, The Decomposition of Water by X-Rays in the Presence of the Iodide or Bromide Ion. *J. Chem. Phys.* **3**, 596 (1935).
37. M. H. Back and N. Miller, Use of Ferrous Sulphate Solutions for X-Ray Dosimetry. *Nature* **179**, 321 (1957).
38. M. Cottin and M. Lefort, Etalonnage absolu du dosimètre au sulfate ferreux. Rayons X mous de 10 et 8 KeV. *J. Chim. Phys.* **53**, 267 (1956).
39. J. L. Haybittle, R. D. Saunders, and A. J. Swallow, X- and γ-Irradiation of Ferrous Sulphate in Dilute Aqueous Solution. *J. Chem. Phys.* **25**, 1213 (1956).
40. V. I. Zatulovskii, N. I. Vitushkin, D. I. Nariadchikov, and V. I. Petrovskii, Istochniki ioniziruiushchikh izluchenii dlia radiatsionno-khimicheskikh issledovanii. *In* " Trudy I Vsesoiuznogo soveschaniia po radiatsionnoi khimii," p. 313. Akademiia Nauk SSSR Mokva, 1958.
41. R. W. Hummell, G. R. Freeman, A. B. Van Cleave, and J. W. T. Spinks, Rotating Sector Methods Applied to Reactions Induced by ^{60}Co Gamma Rays. *Science* **119**, 159 (1954).
42. I. G. Draganić, K. Sehested, and N. W. Holm, Influence of Dose Rate at Large Absorbed Doses on the Oxalic Acid Dosimeter. Danish Atomic Energy Commission, Risö Report 112 (1967).
43. G. Czapski and G. Stein, The Oxidation of Ferrous Ions in Aqueous Solution by Atomic Hydrogen. *J. Phys. Chem.* **63**, 850 (1959).
44. K. Schmidt and E. J. Hart, A Compact Apparatus for Photogeneration of Hydrated Electrons. *Advan. Chem. Ser.* **81**, 267 (1968).
45. A. O. Allen and R. A. Holroyd, Peroxide Yield in the γ-Irradiation of Air-Saturated Water. *J. Amer. Chem. Soc.* **77**, 5852 (1955).
46. Z. D. Draganić, I. G. Draganić, and M. M. Kosanić, Radiolysis of Oxalate Solutions in the Presence of Oxygen Over a Wide Range of Acidities. *J. Phys. Chem.* **68**, 2085 (1964).

In considering chemical changes induced by radiation, we have seen how important it is to know accurate values of radiation-chemical yields for a quantitative interpretation of the process observed. Since these parameters are calculated from the amount of chemical change and energy absorbed (radiation dose), it is evident that the radiation chemist not only needs precise methods of analytical chemistry but also those of radiation dosimetry.

The absorbed radiation dose is the amount of energy absorbed by a unit mass of the material irradiated. A unit of absorbed dose is 1 rad = 100 erg gm^{-1}. Related units are 1 krad = 1000 rad and 1 Mrad = 10^6 rad. In the radiation chemistry of aqueous solutions, the energy absorbed is often expressed in electron volts: 1 rad = 6.24 × 10^{13} eV gm^{-1}.

Different chemical and physical phenomena are used for measuring the absorbed dose; dosimeters are systems used for such measurements. Radiation dosimetry is a wide area of research and the reader interested in its various aspects is referred to the relevant literature [1, 2]. A manual edited by Holm and Berry [3] gives not only the principles of radiation dosimetry and introductions to a number of areas of associated problems but also step-by-step instructions for the use of about twenty selected dosimetry systems.

In this chapter we shall consider only some methods used in the radiation chemistry of aqueous solutions. More detail on this subject may be found in the excellent book of Kabakchvi et al. [4], as well as in some chapters on chemical dosimetry published elsewhere [5–7].

CHAPTER EIGHT

Aqueous Chemical Dosimeters

I. CHEMICAL CHANGE AS A MEASURE OF ABSORBED DOSE

The amount of a chemical change produced by the passage of radiation through a medium depends on the amount of energy absorbed by the medium. Hence, any well-defined radiation-chemical change may in principle serve as a measure of absorbed dose. In practice, however, in order that the system chosen may serve as a routine chemical dosimeter, it must satisfy a number of conditions:

● The amount of chemical change should be linearly proportional to the dose absorbed. For powerful radiation sources (nuclear reactors, accelerators, multikilocurie isotopic sources) it is also often necessary that this proportionality holds in megarad and multimegarad regions.

● Experimental conditions which change in the course of irradiation, such as accumulation of radiolysis products or change in pH, must not affect the reproducibility of dosage curves.

● The amount of chemical change must be independent over a wide range of the LET of radiation, of dose rates, and of temperatures. If this is not the case, the dependence must be well established.

● Chemical dosimeters should be manufactured from commercial chemical-grade substances, without any additional purifications, and the samples should be treated under normal laboratory conditions. For measurements in nuclear reactors, the chemical composition of the substances used must be such that the radioactivity induced is negligible and that it does not interfere with normal handling of the samples irradiated.

● A method that is precise but also simple should be used for determining the chemical change which serves as a measure of absorbed energy.

- In the case where the relation between the dose absorbed and the chemical change measured is not linear, the calculation of the dose must be simple.

It should be recalled that a chemical dosimeter consists of a substance whose radiation-chemical change serves as a measure of absorbed energy, and of a vessel in which the substance is irradiated. The irradiation vessel is an important part of any chemical dosimeter, but this fact is often disregarded. Only a few systematic investigations have been carried out in connection with the effect of the surface of the reaction vessel on radiation-induced chemical changes in aqueous solutions. However, some remarks may be made. It seems that if a solution with a volume from one to several hundred milliliters is irradiated in a glass vessel of a diameter larger than 10 mm, then the size of the vessel and the volume of the solution do not appreciably affect the reaction yield. Plastic vessels are less convenient than those of glass, often only because of difficulties in cleaning. Special care is needed in working with vessels of stainless steel, which are sometimes indispensable in experiments with accelerators and nuclear reactors. In general, the following precaution should be taken. If a chemical dosimeter is used under experimental conditions considerably differing from those under which the yield of its reaction was measured (effective volume of solution, size of the dosimeter vessel, and material of which it is made), it is necessary to check experimentally whether or not these conditions affect the value of radiation yield used in calculating the dose.

Table 8.1 summarizes different radiation-chemical reactions in aqueous solutions which have been proposed for the chemical dosimetry of radiations. It should be noted that none of the systems given in Table 8.1 satisfies all the conditions mentioned above. This is not surprising if one takes into account the complexity of phenomena occurring in irradiated water and ample possibilities offered by various radiation sources. Dose rates vary from a few rad per second to about 10^{14} rad sec^{-1}; the LET of available radiations ranges from 0.02 eV Å^{-1} for ^{60}Co γ-rays up to about 700 eV Å^{-1} for the fission fragments of ^{235}U.

The Fricke dosimeter satisfies most of the conditions mentioned above, but in the kilorad dose range only. The results obtained in different laboratories with systems proposed for the megarad dose range show deviations considerably larger than those due to possible experimental errors of measurement under strictly defined conditions. The quite poor reproducibility was often ascribed to impurities. It seems, however, that the cause should be sought primarily in irradiation conditions which, if not maintained as in the calibration measurement, may affect the secondary reactions and thereby the trend of dosage curves. The mechanisms of reactions induced by large doses (in

TABLE 8.1.

Some Systems Used or Proposed for Chemical Dosimetry in Radiation Chemistry of Water and Aqueous Solutions

System	Dose range, rad	Dosimeter solution	Chemical change and its measurement
Ferrous sulfate (Fricke dosimeter)	1×10^3–4×10^4	Air-saturated 0.4 M H_2SO_4, 1×10^{-3} M $FeSO_4$, 1×10^{-3} M NaCl	Fe^{3+} formation; spectrophotometry
Ferrous–cupric	5×10^4–1×10^6	Oxygen-saturated, 5×10^{-3} M H_2SO_4, 1×10^{-3} M $FeSO_4$, 1×10^{-2} M $CuSO_4$	Fe^{3+} formation; spectrophotometry
Ceric sulfate	1×10^4–2×10^7	Aerated, 0.4 M H_2SO_4, $Ce(SO_4)_2$ from 2×10^{-4} to 5×10^{-2} M	Reduction of Ce^{4+}; spectrophotometric measurements of $[Ce^{4+}]$ before and after irradiation
Oxalic acid	1.4×10^6–1×10^8	Initially aerated solutions of $H_2C_2O_4$, from 5×10^{-2} to 0.6 M	Oxalic acid decomposition; spectrophotometry or titration
Sodium formate	1×10^6–8×10^7	Deaerated H_2O, HCOONa from 5×10^{-2} to 0.3 M	Formate ion decomposition; $KMnO_4$ titration
Benzene	5×10^3–7×10^4	Aerated H_2O, 2×10^{-2} M benzene	Formation of phenol and phenol-like compounds; spectrophotometry
Dyes	10–10^5	Aqueous solutions; various concentrations of methylene blue, methyl orange, indigo carmine	Color degradation; spectrophotometry
Hydrocarbons releasing HCl under the action of radiation	10–10^6	Aqueous solutions; various concentrations of chloroform, trichloroethylene, chloral hydrate	HCl formation; colorimetric measurement of the coloration produced by the reaction of HCl with a suitable indicator

Continued

TABLE 8.1—*Continued*

System	Dose range, rad	Dosimeter solution	Chemical change and its measurement
Optically active hydrocarbons	10^7–10^8	Aqueous solutions; various concentrations of glucose, saccharose, maltose	Decrease in optical activity; polarimetry
Water	1–10^4 rad per pulse	Aqueous solution of ethanol (1×10^{-2} M) or H_2(7×10^{-4} M) and NaOH(10^{-4}–10^{-2} M)	Concentration of hydrated electron formed; fast spectrophotometry
Potassium thiocyanate	10–10^4 rad per pulse	Aqueous solution of 1×10^{-2} M KCNS, saturated with O_2 or N_2O	Concentration of $(CNS)_2^-$ formed; fast spectrophotometry
Potassium ferrocyanide	10^2–10^4 rad per pulse	Aqueous solution of 5×10^{-3} M $K_4Fe(CN)_6$; saturated with O_2 or N_2O	Concentration of ferricyanide formed; fast spectrophotometry

the megarad and multimegarad ranges) are much more complex than in the low dose range and, without exception, are not quantitatively established. The amount of products accumulated under these conditions is not always, as is assumed, of no effect on the reaction mechanism. Numerous studies on aqueous solutions of dyes, or of hydrocarbons which release hydrochloric acid under the action of radiation, are fairly illustrative in this respect.

Accurate knowledge of the radiation yield is very important for the use of a chemical dosimeter. It is acquired in careful experiments where the energy absorbed is precisely measured with a primary dosimeter, most often a calorimeter, or with an already calibrated chemical dosimeter.

Chemical dosimeters are particularly suitable for routine measurements. Their advantage is that they are cheaper and simpler to use than many physical dosimeters. Their other advantage is the possibility of choosing a dosimetric system close in chemical composition to the system studied, hence similar in absorption characteristics to the latter. Any physical or chemical dosimeter measures only the dose absorbed in its effective volume, and in routine work it is desirable that the dosimeter gives a direct measure of the dose absorbed in the sample studied. This is easier to achieve by using chemical rather than physical dosimeters, and aqueous chemical dosimeters are used in the radiation chemistry of aqueous solutions. The calculation of the absorbed dose is usually simpler, since the corrections, if required, concern only a few well-established parameters. It is necessary, however, to be very prudent when considering such corrections. Brynjolfsson [8] drew attention to a significant correction factor to be used in γ-ray dosimetry, which was usually ignored—the ratio of absorbed dose build-up factors for the dosimeter system and the sample studied. This factor is due to the softening of γ rays as they penetrate the irradiated medium and to the fact that the absorption cross sections appreciably increase (particularly in materials with higher Z values) with decreasing photon energy. This difference in absorption characteristics, due to the degraded spectra, is large in the ceric sulfate dosimeter where heavy cerium ions give rise to higher absorbed doses at loci where the spectrum is degraded. If, for example, there is a 16 cm layer of liquid between the ^{60}Co point source and the sample studied, then the dose absorbed in 0.4 M ceric sulfate will be 1.7 times as high as that in water irradiated under the same conditions. This effect is small for the Fricke dosimeter—an increase of 0.2 and 0.8 % at distances of about 16 and 32 cm, respectively. In oxalic acid the difference in absorption is negligible even with substantially degraded spectra (at about 32 cm from a ^{60}Co point source the dose absorbed in 0.1 M $H_2C_2O_4$ solutions is the same as the corresponding dose absorbed in water). It should be noted that in most experiments in radiation-chemical research on water and aqueous solutions only relatively thin layers (less than 5 cm) of dilute solutions (less than 1 M) are irradiated. This means that the irradiation takes place under unperturbed spectral conditions. However, this effect may be important in experiments where higher concentrations of heavier ions are used.

In electron-irradiation dosimetry the spectral degradation problems do not arise.

In experiments with internal radiation sources, absorbed doses are more reliably determined by calculation on the basis of precise data on the radioactivity of the source dissolved and the known energy of its radiation. The same applies to certain irradiations in a nuclear reactor, where the main internal source of energy is a nuclear reaction such as ^{6}Li(n, α)^{3}H. Here, the calculation of absorbed doses is made by using the data on isotope concentration, neutron flux, the energy of the reaction serving as the internal source, and the cross section for the reaction with neutrons. Calculations also may give reliable data in some experiments with accelerator beams. In certain cases calorimeters prove to be the only reliable dosimetric systems: calibration of chemical dosimeters, measurements at very large dose rates obtained by accelerators or in mixed radiation fields such as in a nuclear reactor.

II. Ferrous Sulfate Dosimeter (the Fricke Dosimeter)

We have seen in Chapter One (Section IV) that the oxidation of ferrous ions in acidic aerated solutions was proposed as a chemical dosimeter in the early days of radiation chemistry. After 40 years, the Fricke dosimeter is still the most widely used chemical dosimeter. Numerous studies, reported in several hundred published articles, have contributed to a better understanding and an efficient use of the radiation-induced oxidation of ferrous ions, but the basic idea and the system used remain exactly the same.

The dosimetric solution is made of 1×10^{-3} M $FeSO_4$ (or ferroammonium sulfate) and 1×10^{-3} M NaCl in air-saturated 0.4 M H_2SO_4 (the presence of about 0.25×10^{-3} M O_2). The solutions are prepared from analytical-grade chemicals and water distilled in the same way as for radiation-chemical experiments (Chapter Seven, Section V). If measurements are taken only from time to time, it is necessary to prepare a fresh solution before use. If measurements are carried out every day, or very frequently in the course of the same day, then use can be made of a stock solution. This is a 0.1 or 0.01 M $FeSO_4$ solution in 0.4 M sulfuric acid (in preparing it, ferrous sulfate is dissolved in acid and not in water). It should be kept in normal flasks previously cleaned in the same way as vessels for radiation-chemical experiments. For good stability it is best to keep the stock solution in a refrigerator. In this way the slow oxidation of ferrous ions by oxygen from air, which is particularly troublesome in measurements at the lower limit of doses or below it, is reduced to a minimum. If the chemicals are pure and the conditions of storage those stated above, then this spontaneous oxidation in a 1×10^{-2} M solution is of the order of magnitude of 1 μmole per day.

The ferric ions produced by radiation are most suitably measured spectrophotometrically at 3040 Å. The optical density of ferrous ions is practically negligible at this wavelength. The Lambert–Beer law holds up to 10^{-2} M Fe^{3+}, and the molar extinction coefficient (ε) is 2197 M^{-1} cm^{-1} at 25°C; ε increases by 0.69% for each degree of increase in temperature. These figures [6] represent the mean values of a number of measurements performed in different laboratories.

Absorbed dose, D (in rad), is calculated by the following general formula:

$$D = \frac{N \Delta(OD) \, 100}{\varepsilon 10^3 G(Fe^{3+}) f \rho l} \quad \text{rad} \tag{1}$$

where

N is Avogadro's number, 6.022×10^{23} molecules per mole
$\Delta(OD)$ is the difference between the optical densities of irradiated and control samples
ε is the molar extinction coefficient, 2.197 M^{-1} cm^{-1} at 25°C

II. FERROUS SULFATE DOSIMETER

f is the conversion factor for transition from electron volts per milliliter units into rad, 6.245×10^{13}
ρ is the density of the dosimeter solution, 1.024 for 0.4 M H_2SO_4
l is the optical path length, centimeters
$G(Fe^{3+})$ is the radiation-chemical reaction yield under given conditions.

For 0.4 M H_2SO_4, 1-cm absorption cells, and $G(Fe^{3+}) = 15.6$, Eq. (1) reduces to

$$D = 2.75 \times 10^4 \times \Delta(OD) \quad \text{rad} \qquad (2)$$

assuming that optical density measurements are performed at 25°C. If this is not the case, the above-mentioned temperature correction (0.69% per degree) should also be taken into account.

We know fairly well how the radiation yield of ferric ions, formed under the action of ^{60}Co γ-rays, depends on various experimental conditions. A volume of dosimetric solution, 2 ml $< V <$ 470 ml, was found to have no appreciable effect on the values measured. The vessels used are in most cases made of glass. The concentration of sulfuric acid should be allowed to decrease only down to 0.05 M, since in less acid solutions the reproducibility of the values measured is poorer and the proportionality between the dose absorbed and the amount of ferric ions produced is no longer linear. Ferrous sulfate concentration in the range from 1×10^{-4} to 1×10^{-2} M has no effect on the reaction yield measured. The concentration of oxygen in solution in equilibrium with air is sufficient for doses up to 50 krad. Nevertheless, for the sake of good reproducibility it is desirable that irradiations not be carried out before the complete consumption of oxygen; hence, a dose of 40 krad is taken as the upper limit. After consumption of oxygen, the reaction yield decreases down to 8. The effect of temperature on $G(Fe^{3+})$ is less than 0.1% per degree.

The radiation-chemical yield for the Fricke dosimeter has been determined in numerous experiments, and Table 8.2 gives some of the values recommended [2,6] for use in Eq. (1). For high-LET radiations Fig. 5.4 may be consulted; it shows a general picture of the effect of LET on $G(Fe^{3+})$. Some other details on the LET dependence are also discussed in Chapter Five, Section VI. In the same chapter, Section VII, the effect of the dose rate on the Fricke dosimeter is considered, and Fig. 5.6 shows the $G(Fe^{3+})$ dependence on the dose rate of accelerated electrons. It is worth mentioning here that $G(Fe^{3+}) = 15.5$ can be used at higher dose rates, up to 600 rad per 1.4-μsec pulse. This dose-rate range can be extended to 10,000 rad per 1.4-μsec pulse if the dosimetric solution is slightly modified. It should contain [3] a higher concentration of $FeSO_4$ [or $Fe(NH_4)_2(SO_4)_2$], 1×10^{-2} M, in 0.4 M H_2SO_4. Sodium chloride must be absent. Furthermore, the dosimetric solution must be oxygen-staturated (about 1×10^{-3} $M O_2$). In the dose calculation, $G(Fe^{3+}) = 16.1$ should be taken. It is worth recalling that the dose measured here approximates the mean dose throughout the cell rather than the dose induced by pulse along the light path in kinetic measurements. Other systems are to be used in the latter case (Section V).

TABLE 8.2

Some $G(Fe^{3+})$ Values for the Fricke Dosimeter for Various Radiations[a]

Radiation	Energy, MeV	$G(Fe^{3+})$
Photons	11–30	15.7
	5–10	15.6
	4	15.5
	2	15.4
^{60}Co γ-rays	1.25	15.6
^{137}Cs γ-rays	0.66	15.3
Accelerated electrons	2	15.5
^{32}P β-rays	0.69	15.4

[a] After the ICRU Report [2] and Fricke and Hart [6].

In principle, there is no possibility of extending the LET independence range. As the example of X rays shows, the value 15.5 which is valid for high-energy photons in a very wide energy range cannot be used for 60-keV X rays, where $G(Fe^{3+}) = 13.8$ was determined [9]. This also calls for caution in interpreting chemical yields observed in experiments with spectrally complex X-ray sources.

The only serious disadvantage of the Fricke dosimeter is the narrow dose range, 4–40 krad in routine practice. The upper limit is low for strong radiation sources, which often provide dose rates considerably higher than 40 krad sec^{-1}. Such rapid reaching of the upper limit raises the problem of accurate timing of irradiation and of adequate concentrations of ferrous ions and oxygen. The procedures proposed for shifting the upper limit consist in increasing the concentration of ferrous ions (up to 5×10^{-2} M) and ensuring a sufficient amount of oxygen (by bubbling it through solution during irradiation). The radiation yield is the same as under the standard conditions, but only up to about 40% of conversion of ferrous ions; the dosage curves are not linear thereafter. The blank should also be treated in the same way in order to avoid the error due to increased oxidation caused by impurities or oxygen from air. In such cases, measurements (at moderate dose rates) up to doses of the order of 1 Mrad are possible. A similar upper limit can be reached with deaerated solutions and $G(Fe^{3+}) = 8.0$, where the dosage curves are linear up to about 25% of conversion of ferrous ions. For doses below the lower limit (4 krad), the problem arises of reliable measurements of small amounts of the ferric ions produced, as well as of errors due to the ferrous ion oxidation by agents other than radiation (air–oxygen, impurities). With supersensitive spectrophotometers, long absorption cells (10 cm), and carefully prepared fresh dosimeter solutions, the lower limit can be shifted down to 0.1 krad or less.

III. CERIC SULFATE DOSIMETER

The reduction of ceric ions in acidic aerated solutions increases linearly with increasing dose. This has been proposed for dose measurements in the dose range 10^5–10^7 rad [10] and higher [11]. Simple spectrophotometric measurement of concentration change of ceric ions is a great advantage of this system. However, there are quite large disagreements between published yields of the radiation-chemical reaction that serves as a measure of absorbed dose. The yields as given by different authors vary from 2.04 to 2.65. These variations are accounted for by the effect of impurities on the mechanism of radiolysis or its photosensitivity. We see, however, that even under such well-defined conditions as those in ^{60}Co irradiations, reported yields vary much more than is to be expected from possible errors in chemical analysis and in dosimetry. As was already pointed out (Section I), the main cause of this seems to lie in some not-well-understood experimental conditions which affect certain secondary reactions and thereby the trend of dosage curves. The reaction mechanism at large doses is still not so fully investigated and explained as the reduction of ceric ions occurring in the kilorad dose range. Furthermore, as is seen in Section I, some troubles may be caused by the difference in energy absorption at loci where the spectrum of incident radiation is heavily degraded.

Dosimeter solutions are prepared from ceric sulfate in 0.4 M H_2SO_4. Since absorbed dose is calculated from change in concentration of ceric ions and the error is the larger the smaller the difference, it is necessary to choose the initial concentration with respect to the range of absorbed doses in which measurements are to be carried out. Table 8.3 gives the recommended

TABLE 8.3

INITIAL CONCENTRATIONS OF CERIC SULFATE FOR MEASUREMENTS IN DIFFERENT DOSE REGIONS[a]

Dose range, Mrad	0.01–0.06	0.06–0.5	0.5–4.0	4.0–20.0
Molar concentration of Ce(SO$_4$)$_2$	2×10^{-4}	1.5×10^{-3}	1×10^{-2}	5×10^{-2}

[a] After Bjergbakke [12].

initial concentrations for various dose ranges. The solutions can be prepared from a stock solution (0.1 M) which must be kept in a dark flask. The chemicals and distilled water used must be carefully purified. Since the system is photosensitive, in handling dosimeter samples exposure to light should be avoided as far as possible; the blanks must always be treated under the same conditions. It is assumed that the initial concentration of ceric sulfate in the dosimeter solution, over the range given in Table 8.3, has no effect on measured yields [6] or that the yield decreases by about 13% as the concentration increases from 2×10^{-4} to 5×10^{-2} M [12]. The presence of oxygen in dosimeter solutions does not appreciably affect measured yields. At temperatures higher than 30°C the radiation-chemical yield was found to be somewhat higher, but the data are scarce and in disagreement.

Changes in ceric ion concentration caused by radiation are most conveniently determined by measuring the optical density of Ce^{4+} ions at 3200 Å, where the molar extinction coefficient is 5610 M^{-1} cm^{-1} at 25°C. It is desirable, however, that ε be determined for given experimental conditions. The Lambert–Beer law holds up to 2×10^{-4} M Ce^{4+}. The optical densities are not affected by the optical absorption of Ce^{3+} and temperature changes.

For absorbed dose calculation from measured optical densities use can be made of Eq. (1) by introducing the corresponding values for the ceric system. Since published values of $G(Ce^{3+})$ considerably vary, the average value [6], $G(Ce^{3+}) = 2.4$, should be used with caution. The use of an empirical calibration curve for given experimental conditions is often preferred.

IV. Oxalic Acid Dosimeter

Oxalic acid molecules in aqueous solutions decompose under irradiation to give CO_2 as the main product. The amount of oxalic acid decomposed has been proposed as a measure of the absorbed dose in multimegarad-range dosimetry [13], and possibilities for routine use of oxalic acid in H_2O and D_2O solutions in chemical dosimetry have been described [14,15].

The dosimeter solution is prepared from distilled water and commercial analytical-grade $H_2C_2O_4 \cdot 2H_2O$. Appropriate initial concentrations for measurement in different absorbed dose regions can be obtained from Table 8.4. The system is neither photosensitive nor particularly sensitive to impurities, and a stock solution can be used if it is stored under conditions usual in analytical chemistry. If measurements are carried out only from time to time, it is desirable to prepare fresh solutions. Ampuls made of glass or, for reactor experiments, of silica, serve as dosimeter vessels. In the latter case, it is convenient that the ampuls be sealed. They are filled with solutions in such a

IV. OXALIC ACID DOSIMETER

TABLE 8.4

INITIAL CONCENTRATIONS OF OXALIC ACID FOR MEASUREMENTS IN DIFFERENT DOSE REGIONS[a]

Dose range, Mrad	1.4–8	2.8–16	7–40	14–80	17–100
Molar concentration of oxalic acid	5×10^{-2}	1×10^{-1}	0.25	0.50	0.60

[a] After Marković and Draganić [14].

way that a free volume, about one quarter of the total ampul volume, remains over the solution. Irradiated samples are handled under normal laboratory conditions, in the presence of light and air. Precautions should be taken in opening ampuls exposed to large doses (several tens of megarads), since the gas pressure is high. In such cases, samples should be cooled before opening by plunging in liquid air. For measurements of oxalic acid concentration, both NaOH titration and spectrophotometry with copper–benzidine [16] give satisfactory results. For titration with NaOH, the dosimeter solution is diluted with water, and CO_2 is removed by heating at 80–90°C for half an hour. The titrations are carried out with a standard 0.1 N solution in the presence of phenolphthalein (1% solution) as indicator. Spectrophotometric measurements are performed at 2480 Å. The copper–benzidine reagent is made by mixing equal volumes of solutions A and B prepared in the following way: Solution A: 161 mg of recrystallized benzidine hydrochloride is dissolved in 5 ml of 30% acetic acid and diluted with water to 500 ml in a volumetric flask. Solution B: 375 mg of copper acetate is dissolved in 500 ml of distilled water in a volumetric flask. If the capacity of the volumetric flask in which the dosimeter solution is mixed with the reagent is V, then the amount of the reagent A + B to be added is $V/5$. The molar extinction coefficient is about 2500 M^{-1} cm^{-1} and depends on the purity of the chemicals used. It has a negative temperature coefficient, -0.7% per degree. The optical density is linearly proportional to the oxalic acid concentration in the range 1.4×10^{-5} to 2.5×10^{-4} M.

The decomposition rate of oxalic acid decreases with increasing absorbed dose. It can be represented by an expression for the first-order processes, which in a suitable form gives

$$D = aC_0 \log C_0/C \quad \text{eV ml}^{-1} \tag{3}$$

where C_0 and C are the oxalic acid concentrations (in molecules per milliliter) before and after irradiation, respectively. The proportionality factor a is a constant (in electron volts per molecule) for the given irradiation conditions: initial oxalic acid concentration, LET, dose rate, etc. Table 8.5 presents the

TABLE 8.5

THE PROPORTIONALITY FACTOR a FOR OXALIC
DOSIMETER CALCULATION [EQ. (3)][a,b]

Solvent	a_γ	a_n
H_2O	41.5	58
D_2O	50.5	81

[a] Oxalic acid concentrations of 5×10^{-2}–0.6 M.
[b] After Marković and Draganić [14,15].

data for ^{60}Co γ-rays and neutrons. For absorbed dose calculation in the mixed pile radiation field the value a_{pile} is calculated as

$$\frac{1}{a_{\text{pile}}} = \frac{F_\gamma}{a_\gamma} + \frac{F_n}{a_n} \tag{4}$$

where F_γ and F_n are the fractions of the γ and neutron components in the absorbed dose. Calorimetric measurements [15] show in the core of a heavy-water-moderated pile that $F_\gamma = 0.66$ and 0.80 and $F_n = 0.34$ and 0.20 for H_2O and D_2O solutions, respectively.

It was found that a does not depend on the initial concentration of oxalic acid between 0.05 and 0.60 M, on dose rate up to 2×10^9 rad sec^{-1}, and on irradiation temperature up to 80°C. Dosimeter solutions are initially aerated, oxygen is quickly consumed and it is not renewed. However, a comparison of a values obtained for ^{60}Co γ rays at eight different laboratories shows variations larger than those to be expected (maximum deviations $\pm 10\%$ from the values given in Table 8.5 [14]). As in the case of ceric sulfate, under certain experimental conditions an unidentified factor may affect the dosage curve. It is then necessary to find an empirical calibration curve.

V. Some Other Systems Used in the Radiation Dosimetry of Water and Aqueous Solutions

We have seen in Chapter Three that the hydrated electron has a high optical density (Fig. 3.3). This enables its concentrations to be reliably measured and has been proposed [6,17], for pulsed-beam dosimetry, as a measure of doses ranging from a few rad per pulse to 10 krad per pulse. The dosimeter solution consists of a carefully deaerated aqueous solution containing ethanol (or molecular hydrogen) and sodium hydroxide. These solutes remove OH radicals and H_3O^+ ions, while the disappearance of e_{aq}^- is mainly due to the recombination reaction. A fast spectrophotometric recording

V. SOME OTHER SYSTEMS

system (Chapter Seven, Section II) is to be used for the optical density measurements. The dose absorbed is calculated [17] by

$$D = 9.65 \times 10^8 \frac{OD}{\varepsilon G(e_{aq}^-)l} \quad \text{rad} \tag{5}$$

where OD is the optical density, l is the path length of the absorption cell (in centimeters), ε is the molar extinction coefficient, and $G(e_{aq}^-)$ is the hydrated electron yield. Depending upon the dose per pulse, two systems have been recommended [17]: (a) for doses over 0.1 krad per pulse and pulse lengths less than 5 μsec (System A); (b) for doses less than 0.1 krad per pulse and pulse lengths greater than 5 μsec (System B). Table 8.6 summarizes various data on the hydrated electron dosimeter.

TABLE 8.6

THE HYDRATED ELECTRON DOSIMETER[a]

	System	
Recommendation	A	B
Dose rate, krad per pulse	0.1–10	0.001–0.100
Pulse length, μsec	<5	≥5
Chemical composition	1×10^{-2} M ethanol 1×10^{-4} M NaOH	Hydrogen saturated (7×10^{-4} M) 1×10^{-2} M NaOH
$\varepsilon \times G(e_{aq}^-)$ at 5780 Å	2.94×10^4	7.12×10^4
at 7000 Å	5.0×10^4	12.1×10^4

[a] According to Hart and Fielden [17].

Thiocyanate has also been used for pulsed-beam dosimetery. The species measured is $(CNS)_2^-$, which displays a strong absorption with a peak at 5000 Å and a molar extinction coefficient of 7.1×10^3 M^{-1} cm^{-1}. In Chapter Four, Section III, we have considered the pulse radiolysis of aqueous solutions of thiocyanate and the inconveniences of using this scavenger in competition experiments. However, these disadvantages represent no obstacle in using this system for dose measurements in pulsed-beam experiments. The dosimeter solution consists of a 1×10^{-2} M aerated solution of KCNS. The concentration is not critical and the dosimetric solution is stable [18]. Oxygen removes hydrated electrons and hydrogen atoms, and the product observed arises from solute reaction with the hydroxyl radical only. The dose absorbed is calculated by Eq. (5) given above for the hydrated electron dosimeter, but the yield of e_{aq}^- is replaced by that of the product observed. The value 2.15×10^4

is recommended for $\varepsilon G(P)$ at 5000 Å [18]; this value is doubled by the addition of N_2O to the dosimeter solution. We still need more data on the dose-rate effects on this system which, by its extreme simplicity, seems very promising for pulsed-beam dosimetry. Also, further calibration is needed to improve its accuracy.

Ferrocyanide oxidizes in irradiated aqueous solutions and the concentration of ferricyanide formed has been proposed as a measure of the dose absorbed. Oxidations involving H_2O_2 and HO_2 (or O_2^-) are slow compared to the OH oxidation step in pulsed-beam experiments (even with a 100-μsec pulse-length) and only $Fe(CN)_6^{3-}$ from hydroxyl radical reaction is measured. These measurements are carried out at 4200 Å and a molar extinction coefficient of $1 \times 10^3 \ M^{-1} \ cm^{-1}$ is used for the dose calculation. If the dosimeter solution consists of $5 \times 10^{-3} \ M \ K_4Fe(CN)_6$ saturated with oxygen, then the absorption displayed at 4200 Å is measured 10 μsec after the pulse and $\varepsilon G(P) = 3.2 \times 10^3$ is used in Eq. (5) for the dose calculation [18]. This value is doubled by the addition of nitrous oxide to the solution. In this case, before irradiation the solution is not saturated with oxygen but only shaken for about 5 min with 5 ml of air per 100 ml of dosimeter solution [19]. The role of oxygen is to remove H atoms and to prevent their reaction with the ferricyanide formed. Since the photochemical oxidation in these measurements might also be appreciable, a filter that cuts wavelengths shorter than 3000 Å should be placed between the source of analyzing light and the irradiation cell (Fig. 7.3) to minimize this effect.

For the million rad range, 0.5–12 Mrad per pulse, the use of an adiabatic calorimeter has been described [20]. The calorimeter body is a thin graphite disk; the output of the attached thermocouple is measured with a fast-response recorder (0.5-sec response time). The graphite disks are of the same size and geometry as the liquid samples for which the dosimetry is designed. The electron stopping power of graphite is very close to that of water.

For measurements of absorbed dose of mixed-pile radiation in the radiation chemistry of water and aqueous solutions, a differential calorimeter has been described [21]. It has three bodies and allows the contributions of two main components of pile radiation to be simultaneously measured; one of the bodies may contain the chemical system studied, which is subsequently chemically analyzed.

References

1. F. H. Attix and W. C. Roesch, eds., "Radiation Dosimetry," Vols. 1–2. Academic Press, New York, 1966.
2. ICRU, "Radiation Dosimetry: X Rays and Gamma Rays with Maximum Photo Energies between 0.6 and 50 MeV." Int. Comm. Radiat. Units Meas., ICRU Rep. 14. Washington, D.C., 1969.

REFERENCES

3. N. W. Holm and R. J. Berry, eds., "Manual on Radiation Dosimetry." Dekker, New York, 1970.
4. A. M. Kabakchi, Ia.I. Lavrentovich, and V. V. Penkovskii, "Khimicheskaia dozimetriia ioniziruiushchih izluchenii." Akad. Nauk Ukrainskoi SSR, Kiev, 1963.
5. N. Miller, Introduction à la dosimétrie des radiations. In "Actions chimiques et biologiques des radiations" (M. Haissinsky, ed.), Vol. 2, p. 144. Masson, Paris, 1956.
6. H. Fricke and E. J. Hart, Chemical Dosimetry. In "Radiation Dosimetry" (F. H. Attix and W. C. Roesch,eds.), Vol. 2, p. 167. Academic Press, New York, 1966.
7. I. Draganić, B. Radak, and V. Marković, Chemical Dosimeters. In "Manual on In-Pile Dosimetry" (A. W. Boyd, ed.). Int. At. Energy Ag. Vienna, 1971.
8. A. Brynjolfsson, A Significant Factor in Gamma Ray Dosimetry. *Advan. Chem. Ser.* **81**, 550 (1968).
9. J. Novotný and Z. Spurný, G (Fe^{3+}) Yield of the Fricke-Miller Ferrosulphate Dosimeter for 60 keV X-Rays from an X-Ray Tube. *Int. J. Appl. Radiat. Isotopes* **19**, 403 (1968).
10. S. I. Taimuty, L. H. Towle, and D. L. Peterson, Ceric Dosimetry: Routine Use at 10^5–10^7 Rads. *Nucleonics* **17**, No. 8, 103 (1959).
11. J. T. Harlan and E. J. Hart, Ceric Dosimetry: Accurate Measurements at 10^8 Rads. *Nucleonics* **17**, No. 8, 102 (1959).
12. E. Bjerbakke, The Ceric Sulfate Dosimeter. In "Manual on Radiation Dosimetry" (N. W. Holm and R. J. Berry, eds.), p. 319. Dekker, New York, 1970.
13a. I. G. Draganić, Action des Rayonnements ionisants sur les solutions aqueuses d'acide oxalique: Acide oxalique aqueux utilisé comme dosimetre chimique pour les doses entre 1,6 et 160 Mrads. *J. Chim. Phys.* **56**, 9 (1959).
13b. I. Draganić, Oxalic Acid: The Only Aqueous Dosimeter for In-Pile Use. *Nucleonics* **21**, No. 2, 33 (1963).
14. V. Marković and I. Draganić, New Possibilities for Routine Use of Oxalic Acid Solutions in Multimegarad Gamma Radiation Dosimetry. *Radiat. Res.* **35**, 587 (1968).
15. V. Marković and I. Draganić, New Possibilities for Routine Use of Oxalic Acid Solutions in in-Pile Dosimetry. *Radiat. Res.* **36**, 588 (1968).
16. Z. D. Draganić, The Spectrophotometric Determination of Some Organic Acids with Copper Benzidine, *Anal. Chim. Acta.* **28**, 394 (1963).
17. E. J. Hart and E. M. Fielden, The Hydrated Electron Dosimeter, In "Manual on Radiation Dosimetry" (N. W. Holm and R. J. Berry, eds.), p. 331. Dekker, New York, 1970.
18. E. M. Fielden and N. W. Holm, Dosimetry in Accelerator Research and Processing. In "Manual on Radiation Dosimetry" (N. W. Holm and R. J. Berry, eds.), p. 261 Dekker, New York, 1970.
19. P. Pagsberg, H. Christensen, J. Rabani, G. Nilsson, J. Fenger, and S. O. Nielsen, Far-Ultraviolet Spectra of Hydrogen and Hydroxyl Radicals from Pulse Radiolysis of Aqueous Solutions. Direct Measurements of the Rate of H + H. *J. Chys. Chem.* **73**, 1029 (1969).
20. C. Willis, O. A. Miller, A. E. Rothwell, and A. W. Boyd, The Dosimetry of Very High-Intensity Pulsed Electron Sources Used for Radiation Chemistry: Dosimetry for Liquid Samples. *Radiat. Res.* **35**, 428 (1968).
21. I. G. Draganić, B. B. Radak, and V. M. Marković, Measurement of Absorbed Dose of Mixed-Pile Radiation in Aqueous Radiation Chemistry. *Int. J. Appl. Radiat. Isotopes* **16**, 145 (1965).

Numbers in parentheses are reference numbers and indicate that an author's work is referred to, although his name is not cited in the text. Numbers in italics show the page on which the complete reference is listed.

Author Index

A

Adams, G. E., 90(3), 94(25), 95(32), 98(45, 47), 99(25, 45, 47, 51, 52a), 100(64), 102 (52a), 107(51, 52a, 64), 110(47), *116*, *117*, *118*, *119*, 130(45, 46), 164

Alfassi, Z. B., 56(67), *84*

Allan, J. T., 14, *19*, 46(8), 73(8), *82*

Allen, A. O., 2, 10(5, 7), 11(64, 65), 14, *15*, *16*, *18*, *19*, 27(8, 10), 28(10), 42, 46(7, 9), 56(9), 61, 62(81), 63, 64(87), 70(87, 102), 71(87), *82*, *85*, *86*, 94(31), 101, 109(82), 110(82), 111(82), 113(31), *118*, *119*, *120*, 122(1), 126, 128(21), 132(1), 135, 136(58), 139(66), 140(57, 58), 141(57), 151, 152, 153, *162*, *163*, *164*, 183, *189*, 199(27, 28), 200, 202, 203(45), *207*, *208*

Anbar, M., 15(98), *19*, 42(81), *45*, 46(24a, 24b), 53, 54(42, 49, 56), 55(49), 56(42, 67, 72, 73, 78), 62(40), 69, 72(124), 75(133), 78, *82*, *83*, *84*, *85*, *86*, *87*, *88*, 94(23), 105 (71), 111(91), 113(98), 115(91), *117*, *120*, *121*, 122(6), 145, 158(116), *162*, *165*, *167*

Ander, S. M., 35(40), 40(40), *44*

Anderson, A. R., 76(137), *88*, 154(92), 155, 156(109), 157, *166*, *167*, 174(13), *189*

Appleby, A., 74, *87*, 150(88), 155, *166*

Arakawa, E. T., 33(24), *43*

Armstrong, D. A., 14, *19*, 145(77), 146, 148(77), *165*

Arthur, J. C., Jr., 109(83), *120*

Ashkin, J., 22(1), 26, 28(1), *42*

Asmus, K. D., 55(66), 64(66, 91), 72(91), *84*, *86*, 97(36), 100(36), *118*, 147(84), 148(84), 149, 150(84), *166*

Attix, F. H., 210(1), *224*

B

Back, M. H., 201(37), *208*

Backhurst, J. D., 139, *165*

Bains, M. S., 109(83), *120*

Barr, N. F., 14, *19*, 46(7), 63, *82*, 109(78), *120*, 198(23), *207*

Barret, J., 33(25), *43*

Bartoníček, B., 147(83b), 148(83b), *166*

Basco, N., 81(148), *88*

Baxendale, J. H., 3(26), 14, 15(26), *16*, *19*, 33(25), 40(57), *43*, *44*, 46(5, 15), 49(29), 54(50), 55(50, 61, 64), 64(50, 88, 89), 70 (110, 115), 71, 72(88, 89), *82*, *83*, *84*, *85*, *87*, 91, 94(20, 21, 28), 99(20, 28, 61), 100(61), 106(61), 107(28), 108, *117*, *118*, *119*, 138, *165*

Bednář, J., 29, 31, 32(15a, 15c), *42*, *43*, 147 (83b), 148(83b), 152, *166*

Belford, G. G., 170(3), 174, 176, 177, 187, *188*, *189*

Bensasson, R. V., 15(99), *19*, 81, *88*, 181, *189*

Benson, S. E., 57(79), *85*

Bernstein, W., 27(9), *42*

Berry, R. J., 210, *225*

Bethe, H. A., 22(1), 26, 28(1), *42*
Bevan, P. L. T., 64(89), 72(89), *85*, 94(28), 99(28), 107(28), *118*
Bielski, B. H. J., 97(42), 98, 109(80, 82), 110(42, 82), 111(82), *118*, *120*, *121*, 135, 140(57), 141(57), 143(71), *164*, *165*
Birkhoff, R. D., 33(24), *43*
Bjerbakke, E., 219, *225*
Blok, J., 106(75), *120*
Boag, J. W., 14, *19*, 46(12, 13, 22), 50, 51, *82*, 94(25), 98(45, 47), 99(25, 45, 47, 51, 52a), 100(64), 102(52a), 107(51, 52a, 64), 110(47), *117*, *118*, *119*, 194(6), 196(13), *206*, *207*
Bonner, O. D., 126(8), *162*
Boyd, A. W., 157(111), *167*, 224(20), *225*
Boyle, J. W., 34(32), *43*, 51(37), 52, *83*, 93(18), 94, *117*, 139(67), 140(67), 141, 147, 148(67), *165*
Braams, R., 56(74), *85*
Bradley, D. L., 145(79), *165*
Breger, A. H., 24(7), *42*, 190, *206*
Bregman-Reisler, H., 56(67), *84*
Bronskill, M. J., 15(100), *20*, 36(48), *44*, 51(36), *83*, 194(7), 197, *206*
Brown, D. M., 55(63), *84*
Brownscombe, E. R., 6, 7, *17*, *18*
Brynjolfsson, A., 215, *225*
Buck, W. L., 48, *83*, 197(19), *207*
Bühler, R. E., 97(34), 100(34), *118*
Burch, P. R. J., 178, *189*
Burchill, C. E., 128(39), *163*
Burton, M., 2, 10, 11(63), *15*, *16*, *18*, 36, *44*, 70, *86*, 160(121, 122, 124), 161(121, 124), *168*, 177, *189*
Buxton, G. V., 54(43), *83*, 102(52b), *119*, 122(7), 128(38), *162*, *163*

C

Caffrey, J. M., 128(26), 155(26), *163*, 177(21), *189*
Caldin, E. F., 15(104), *20*
Cameron, A. T., 4, 9(35), *17*
Capellos, C., 54(50), 55(50), 64(50), *84*
Casey, R., 137(60), 140(60), *165*
Čerček, B., 35(39), 40(58), *44*, 54(44), 55(44, 65), 56(44), *83*, *84*, 94(24), 97(36), 100(36), *117*, *118*, 159(119), *167*, *168*

Chambers, K. W., 56(68), *84*, 97, *118*
Chatterjee, A., 38(51), *44*
Cheek, C. H., 129(42), 143, *164*, *165*
Chevrel, J., 200(30), *207*
Chouraqui, M., 73(130), 74, 76(130), *87*, 128(34), *163*
Christensen, H. C. , 40(66), *45*, 51(35), 70(116, 126), 71(116), *83*, *87*, 93(19), 94(19), 99(19), 113(19), *117*, 197, *207*, 224(19), *225*
Clark, G. L., 7, *18*
Clark, R. W., 2(2), *15*
Coatsworth, K., 147, 148(83a), *166*
Coe, W. S., 7, *18*
Collinson, E., 14, *19*, 46(11), 56(68), 60, *82*, *84*, 97(39, 40, 41), *118*, 145(77), 146(77), 147(83a), 148(77, 83a), 154(94, 95), *165*, *166*, 200(35), *208*
Cottin, M., 33(28), *43*, 201(38), *208*
Coyle, P. J., 47(27), *83*
Currant, J., 99(52a), 100(64), 102(52a), 107(52a, 64), *119*
Czapski, G., 14, *19*, 34, *44*, 46(10), 54(48), 58, 59, 64(92, 93), 65(93), 70(92, 100, 101, 107), 71(100, 121), 77, *82*, *84*, *86*, *87*, 90(4), 93(4), 97(42), 98(46, 48), 109(84, 85), 110(42, 46), *116*, *118*, *119*, *120*, 122(4), 131, 141, 143(71), 144, *162*, *164*, *165*, 202, *208*

D

Dainton, F. S., 13, 14, *18*, *19*, 40(76), *45*, 46(11, 21), 47(27), 54(43), 55(63), 56(68), 60(11), 64(83), 70(83), 71(83, 122), *82*, *83*, *84*, *85*, *87*, 97(39, 40, 41), 99(63), 102(63), 106(73b), *118*, *119*, *120*, 127(17), 128(38, 39), 145(77), 146(77), 147(83a), 148(77, 83a), 154(93, 94, 95), *162*, *163*, *165*, *166*, 200(35), *208*
Daniels, M., 129(43, 44), *164*
Davies, J. V., 54(50), 55(50), 64(50), *84*
Davis, C. M., Jr., 145(79), *165*
Davis, T. W., 11(65), *18*
Day, M. J., 13, *18*, 46(4), *82*
Debierne, A., 4, *17*
DeHeer, F. J., 22(3), *42*
DeMaeyer, L., 40(79), *45*

Desalos, J., 79, *88*
Dixon, R. S., 70(110), *87*
Dobó, J., 3(24, 25), *16*
Donaldson, D. M., 14, *19*, 90(2), *116*
Dorfman, L. M., 2(20), 15(20), *16*, *19*, 40 (59, 62), *45*, 54(55), 55(55, 60), 67, 71(60), 72(125), 80, *84*, *87*, 96(33), 97(35, 37), 98 (46, 49, 50), 99(49), 100(34, 35, 37, 50), 110(46), *118*, *119*, 131, 143(72), *164*, *165*, 194(5), 196, *206*
Draganić, I. G., 34(34), *43*, 54(50), 56(76), 64(51), 71(51), *84*, *85*, 114, 115(100), 116 (100), *121*, 127(14), 128(14, 31a, 31b, 35), 130(14), 134(14), 139(31a, 31b, 35), 140 (14, 31a, 31b, 35), 142(14), 143(35), 148 (85), *162*, *163*, *166*, 174(15), 187(27), *189*, 192(3), 193(4), 198(22, 26), 202(42), 206 (46), *206*, *207*, *208*, 210(7), 220(13a, 13b, 14), 221, 222, 224(21), *225*
Draganić, Z. D., 34(34), *43*, 105, 114, 115 (100), 116(100), *120*, *121*, 127(14), 128 (14, 31a, 31b, 35), 130(14, 49), 134(14), 139(31a, 31b, 35), 140(14, 31a, 31b, 35), 142(14), 143(35), 146, 147(82), 148(82,85), 149, *162*, *163*, *164*, *166*, 174(15), 187(27), *189*, 206(46), *208*, 221(16), *225*
Duane, W., 4, *17*
Dyne, P. J., 175, *189*

E

Ebert, M., 2, 3(26, 31), 15(26), *16*, 46(15), 54 (50), 55(50, 65), 56(77), 64(50, 91), 72(91), *82*, *84*, *85*, *86*, 94(24), 97(36), 100(36, 65), 106(65), 107(65), *117*, *118*, *119*, 159, *167*
Ehrenberg, L., 198(21), *207*
Eigen, M., 40(79), *45*
Eisenberg, D., 126(10), *162*
Ertl, G., 40(80), *45*

F

Failla, G., 29(18), *43*
Faraggi, M., 79, *88*
Farhataziz, 70, *86*, 94(27), *118*, 160(121, 122, 123, 124, 125), 161(121, 124, 125), *168*

Felix, W. D., 40(62), *45*, 55(60), 71(60), *84*, 98, 99(49), *119*
Fendler, J. H., 55(66), 64(66), *84*, 147(84), 148(84), 149, 150(84), *166*
Fenger, J., 40(66), *45*, 70(116), 71(116), *87*, 93(19), 94(19), 99(19), 113(19), *117*, 224 (19), *225*
Ferrandini, C., 113(95), *121*, 129(41), *164*
Field, F. H., 34(37), *44*
Fielden, E. M., 40(57, 61, 63), *44*, *45*, 51(34), 52(34, 38), 53(39), 54(50, 59), 55(50, 61, 64), 56(78), 64(50), 71(115), *83*, *84*, *85*, *87*, 94 (20, 21), 99(20), *117*, 133, 138(54), 140 (54), 147, 148(55), 149, 150(86), *164*, *166*, 174(12, 14), *189*, 222(17), 223(17,18),224, (18),*225*
Fiquet-Fayard, F., 29, 30(14a), *42*
Flanders, D. A., 128(37), *133*, *134*, *135*, *163*, 174, *189*
Florin, R. E., 93(13, 14), *117*
Fowles, P., 71(122), *87*
Francis, J. M., 54(50), 55(50), 64(50), *84*
Franklin, J. L., 34(37), *44*
Freeman, G. R., 42(82), *45*, 127(16), 128, *162*, 201(41), *208*
Fricke, H., 5, 6, 7, *17*, *18*, 40(67), *45*, 64(82), 70(82), 71(82), *85*, 99(58),110(86), 112(58), *119*, *120*, 128(37), 133, 134, 135, 155(108), 156, *163*, *167*, 174(8, 11, 16,17), *188*, *189*, 201(36), *208*, 210(6), 217(6), *218*, 220(6), *225*
Friedman, H. L., 70(104), *86*

G

Gall, B. L., 40(62), *45*, 55(60), *84*, 98(49), 99(49, 50), 100(50), *119*, 143(72), *165*
Ganguly, A. K., 174, 177, *188*
Gerischer, H., 40(80), *45*
Getoff, N., 33(31), *43*
Ghormley, J. A., 11(65), *18*, 34(32), *43*, 51 (37), 52(37), 76(136), *83*, *88*, 93(18), 94 (18), *117*, 128(22), 157, *163*, *167*
Giesel, F., 3, *17*
Gilbert, C. W., 54(50), 55(50), 64(50), *84*, 94(24), *117*
Glazunov, P. Ya., 155, 157(112, 113, 114), *167*
Goodman, C. D., 153(91b), *166*

Gordon, S., 40(53, 54, 63, 75), *44*, *45*, 52 (38), 54(45, 46, 47, 53), 55(46, 47, 53), 56 (45, 46, 70), 57(53), 64(46, 53), 71(46), *83*, *84*, *85*, 93(17), 99(17, 55), 113(17), *117*, *119*, 131(52), 153(90), *164*, *166*
Gould, R. F., 3(27, 29), 15(27, 29), *16*, 46 (16), *82*
Gray, P., 90, *116*
Green, B. C., 109(81), *120*
Greenshields, H., 15(96), *19*, 46(17), *82*
Greenstock, C. L., 56(69), *85*, 107(77), *120*
Gross, W., 29(18), *43*

H

Habersbergerova, A., 15(97), *19*
Haissinsky, M., 2(13), *16*, 71(120), *87*, 122 (2), 136, 140(59), 144, *162*, *164*
Halpern, J., 71(119), *87*
Hamill, W. H., 29(13), *42*(83), *45*
Hardwick, T. J., 145(76), 147(76), 148(76), *165*, 200(31, 33), *208*
Harlan, J. T., 219(11), *225*
Hart, E. J., 2, 6, 7(59),13,14,*15*,*16*,*18*,*19*, 36 (44), 40(53, 54, 55, 61, 63, 75), *44*, 45, 46 (12, 13, 18, 25, 26), 51(34, 35), 52(38), 53 (39), 54(42, 45, 46, 47, 49, 53, 54, 59), 55 (46, 47, 49, 53), 56(42, 45, 46, 70, 71, 72, 73, 78), 57(53), 64(46, 53), 70(109), 71(46), 72(71), 76(137), *82*, *83*, *84*, *85*, *86*, *88*, 90 (1), 93(17), 99(17, 55), 113(17), *116*, *117*, *119*,127(15),128(15,27),130(15,46,47),131 (52), 133, 134, 138(54), 140(15, 54), 142 (15), 147, 148(55), 149, 150(86), 153(90), 154(92), 155(108), 156(109), 157, 158(116), *162*, *163*, *164*, *166*, *167*, 174(12, 13, 14), *189*, 196, 198(20), 200(34), 201(36), 202, 205(10), *207*, *208*, 210(6), 217(6), 218, 219 (11), 220(6), 222(17), *225*
Haybittle, J. L., 201(39), *208*
Hayon, E., 14, *19*, 46(6, 9), 56(9), 60(6), 61, 62, 70(108), 71(108), 76(139), *82*, *86*, *88*, 113(94), *121*, 122(3), 128(3, 32, 36a, 36b), 139(64), 143(32), 144, 146(81), 148(81), *162*, *163*, *165*, *166*
Hedvig, P., 3(25), *16*
Helman, W. P., 22(6), *42*
Henglein, A., 54(41), 55(62), 64(91), 72(91), *83*, *84*, *86*, 97(36), 100(36), *118*

Hentz, R. R., 70(103), *86*, 160, 161(121, 124, 125), *168*
Hickel, B., 40(60), *45*, 54(58), 56(58), 64(58), *84*
Hinojosa, O., 109(83), *120*
Hochanadel, C. J., 11(65), 13, *18*, 34(32), 40 (70), *43*, *45*, 51(37), 52(37), 71(118), 76 (136), *83*, *87*, *88*, 92(10), 93(18), 94(18), 99(9), 102(9), *117*, 127, 128(22), 137(60), 140(60), 153(91b), 157, *162*, *163*, *165*, *166*, *167*
Holm, N. W., 202(42), *208*, 210, 223(18), *225*
Holroyd, R. A., 94(31), 113(31), *118*, 128 (21), *163*, 203(45), *208*
Horne, R. A., 126(9), *162*
Howard, A., 2, 3(31), *16*
Hughes, G., 14, *19*, 46(5), *82*, 102(69), 113 (97), *120*, *121*, 128(29), *163*
Hummel, A., 101, *119*
Hummell, R. W., 201, *208*
Hunt, J. W., 15(100), *20*, 36(48), *44*, 51(36), 56(69), *83*, *85*, 107(77), *120*, 194(7), 195 (8), 197(8), *206*

J

Janovský, I., 15, *19*, 147(83b), 148(83b), *166*
Johannin-Gilles, A., 29(16), *43*
Johnson, C. G., Jr., 70(103a), *86*, 160(126), *168*
Johnson, G. R. A., 3(28), *16*, 139(68), *165*
Johnson, K., 195(9), 197(9), *207*
Jortner, J., 49(30), 50(32), 64(93, 94, 96), 65 (93, 96), 68,70(100), 71(100,121), *83*,*86*,*87*
Julien, R., 200(30), *207*

K

Kabakchi, A. M., 210, *225*
Kabakchi, S. A., 2, *16*
Kailan, A., 4, *17*
Katakis, D., 34, *44*, 70(102), *86*
Kauzmann, W., 126(10), *162*
Keene, J. P., 3(26), 14, 15(26), *16*, *19*, 40 (56, 57), *44*, 46(14, 15), 51, 54(50, 57), 55 (50, 57, 61, 64), 64(50, 85, 91),71(85,115) 72(91), *82*, *84*, *85*, *86*, *87*, 94(20, 21), 99 (20), *117*, *118*, 130(48), 131(48), 155, *164*, *167*, 196(15), 197(15), *207*

AUTHOR INDEX

Kennedy, J. M., 175, *189*
Kenney, G. A., 81(148), *88*
Kernbaum, M., 4(39), *17*
Khan, A. A., 99(61), 100(61), 106(61), *119*, 138, *165*
King, C. G., 109(78), *120*
Kistemaker, J., 22(3), *42*
Klenert, M., 196(12), *207*
Klippert, T., 195(9), 197(9), *207*
Knight, R. J., 160(127), *168*
Kochanny, G. L., Jr., 153(91b), *166*
Kolotyrkin, V. M., 93, *117*
Kongshaug, M., 35(39), *44*
Kosanić, M. M., 105(70),*120*,128(31a,31b), 139(31a, 31b), 140(31a, 31b), *163*, 206 (46), *208*
Kraljić, I., 92(11), 106(73a), *117*, *120*
Kroh, J., 109, *120*, 154(93, 94, 95), *166*, 200 (35), *208*
Kuppermann, A., 115(101), 116(101), *121*, 134(56), 135, *164*, 170, 172(2), 174, 176, 177,181, 182, 184, 185, 186, 187, *188*, *189*
Kurien, K. C., 177, *189*

L

Laidler, K. J., 159(120), *168*
Lampe, F. W., 34, *44*
Land, E. J., 54(50), 55(50), 56(77), 64(50), *84*, *85*
Lanning, F. C., 9, *18*
Lavrentovich, Ia. I., 210(4), *225*
Lea, D. E., 13, *18*, 92(8), *117*
Lefort, M., 201(38), *208*
Levanon, H., 109(84), *120*
Lewis, D., 29(13), *42*
Lifschitz, C., 73(128), *87*, 145, *165*
Lind, S. C., 9, *18*
Linnenbom, V. J., 129(42), 143(73), *164*, *165*
Logan, S. R., 47(27), *83*, 127(17), 128, *162*
Loman, H., 106(75), *120*
Losee, J. P., 139(66), 140(66), *165*

M

McCarthy, R. L., 15, *19*
McDonell, W. R., 198(20), *207*

MacLachlan, A., 15, *19*
McNesby, J. R., 33(29), *43*
Magee, J. L., 2, 13, *16*, *18*, 35, 36, 37, 38(51), 44, 46(1, 2), 73(129), *81*, *87*, 115(102), *121*, 170, 171(6), 174, 177, 180, *188*, *189*
Mahlman, H. A., 73(127), 76(140, 141, 142, 145), 79, 80(140), *87*, *88*, 128(25, 28a, 28b), 139(28a, 28b), 140(28a, 28b, 65, 67), 147, 148(67), *163*, *165*
Makada, H. A., 102(69), *120*
Marketos, D. G., 106, *120*
Marković, V. (M.), 210(7), 220(14, 15), 224 (21), *225*
Masanet, J., 33(28, 30), *43*
Matheson, M. S., 2(20), 15(20, 101), *16*, *19*, 20, 40(52, 53, 54, 72, 75), *44*, *45*, 51(33), 54(46, 52, 53), 55(46, 52, 53), 56(33, 46), 57(53), 64(33, 46, 52, 53), 68, 71(46, 52), *83*, *84*, 93(17), 94(26), 99(17, 26, 54, 55, 56, 57), 100(57), 102, 112(56, 57), 113(17), *117*, *118*, *119*, 194(5), 196(16, 17), *206*, *207*
Matthews, R. W., 105, *120*
Meissner, G., 55(62), *84*
Melton, C. E., 33(22), *43*
Meyerstein, D., 40(68), *45*, 70(114), 71(114), 72(124), 75(133), 78, *87*, *88*, 145, *165*
Michael, B. D., 56(71), 72(71), *85*, 94(25), 98(45, 47), 99(25, 45, 47, 51, 52a), 100 (64), 102(52a), 107(51, 52a, 64), 110(47), *117*, *118*, *119*
Mićić, O., 54(51), 56(76), 64(51), 71(51), *84*, *85*, 146(82), 147(82), 148(82), 149(82), *166*
Milašin, N. J., 192(3), 198(26), *206*, *207*
Miller, N., 14(83),*19*,90(2),*116*,200(32), 201 (37), *208*, 210(5), *225*
Miller, O. A., 157(111), *167*, 224(20), *225*
Milner, D. J., 70, *86*, 160(121, 122, 123, 124, 125), 161(121, 124, 125), *168*
Moreau, M., 70(108), 71(108), 75(132, 135), 76(139), *86*, *88*, 128(32), 139(64), 143(32), *163*, *165*
Morse, S., 6, *17*
Mozumder, A., 22(6), 36(45, 46, 49), 37, 38, *42*, *44*, 115(102), *121*, 171(6), 180, *188*, *189*
Mulac, W. A., 51(33), 56(33), 64(33), *83*, 94 (26), 99(26), *118*, 196(17), *207*
Muller, J. C., 113(95), *121*

N

Nariadchikov, D. I., 201(40), *208*
Navon, G., 64(86, 95), 65(86), 70(86, 95, 105), 71(86), *85*, *86*
Nehari, S., 71(117), 72(117), 73(117), *87*, 138(63), 140(63), *165*
Némethy, G., 145(78), *165*
Nenadović, M. T., 105, *120*, 127(14), 128(14), 130(14, 49), 134(14), 140(14), 142(14), 146(82), 147(82), 148(82, 85), 149(82), *162*, *164*, *166*, 174(15), *189*
Neta, P., 15(98), *19*, 53, 62(40), 69, 72(124, 125), *83*, *86*, *87*, 97(35, 37), 100(35, 37), 105(71), *118*, *120*
Ng, M., 56(69), *85*, 107(77), *120*
Nielsen, S. O., 40(66), *45*, 51(35), 70(116, 126), 71(116), *83*, *87*, 93(19), 94(19), 99(19), 110(88), 111(88), 113(19), *117*, *121*, 131(51), *164*, 197(18), *207*, 224(19), *225*
Niemann, E. G., 196(12), *207*
Nilsson, G., 40(66), *45*, 51(35), 70(116, 126), 71(116), *83*, *87*, 93(19), 94(19), 99(19), 113(19), *117*, 197(18), *207*, 224(19), *225*
Nosworthy, J. M., 54(50), 55(50), 64(50), *84*
Novotný, J., 218(9), *225*
Noyes, R. M., 49(30, 31), *83*
Nurnberger, C. E., 9, *18*

O

Okabe, H., 33(29), *43*
Oldenberg, O., 93(16), *117*
Onsager, L., 34(38), *44*
Ore, A., 22(4), *42*
Ottolenghi, M., 65(96), *86*

P

Pagsberg, P., 40(66), *45*, 51(35), 70(116, 126), 71(116), *83*, *87*, 93(19), 94, 99(19), 113(19), *117*, 197(18), *207*, 224(19), *225*
Painter, L. R., 33(24), *43*
Pecht, I., 113(98), *121*
Peled, E., 54(48), 77, *84*, 141, *165*
Penkovskii, V. V., 210(4), *225*

Petersen, B. W., 5, *17*
Peterson, D. L., 219(10), *225*
Petrovskii, V. I., 201(40), *208*
Phillips, D. L., 174(17), *189*
Pickett, L. W., 7, *18*
Pikaev, A. K., 2(16, 19), 15(19), *16*, 155, 157, *167*, 196, *207*
Platzman, R. L., 2, 13, *16*, *18*, 22(2), 29, 30 (2, 20), 31, 32(2), 36(43, 44), *42*, *44*, 46(3), *81*, 145, *165*
Porter, G., 15(102, 105), *20*, 196(14), *207*
Proskurnin, M. A., 93, *117*
Pshezhetskii, S., Ia., 14(81), *19*
Pucheault, J., 113(95), *121*, 122(5), 129(41), 152, 155, *162*, *164*, *166*, 198(24), 200, *207*
Pukies, J., 54(41), *83*

R

Rabani, J., 40(52, 53, 54, 66, 68, 69, 72, 75, 78), *44*, *45*, 51(33), 54(46, 52, 53), 55(46, 52, 53), 56(33, 46), 57(53), 64(33, 46, 52, 53, 90, 94), 65(96, 97), 68, 69(97), 70(90, 107, 114, 116, 126), 71(46, 52, 90, 97, 114, 116, 117, 119), 72(90, 117, 123), 73(117), *83*, *84*, *85*, *86*, *87*, 90(5), 93(17, 19), 94(19, 26), 99(5, 17, 19, 26, 53, 54, 55, 56, 57, 62), 102(5, 53, 62), 110(88), 111(88), 112(56, 57), 113(17, 19), *116*, *117*, *118*, *119*, *121*, 131(51, 53), 138(63), 140(63), *164*, *165*, 196(16, 17), *207*, 224(19), *225*
Radak, B. (B.), 193(4), 198(26), *206*, *207*, 210(7), 224(21), *225*
Radosavljević, B. V., 192(3), *206*
Raef, Y., 64(85), 71(85), *85*, 130(48), 131(48), *164*
Rafi, A., 94(22), 113(22), *117*, 128(30), 140(30), *163*
Ramler, W. J., 195(9), 197(9), 200(29), *207*
Ramsay, W., 4, 9(35), *17*
Rasmussen, O. L., 110(86), *120*
Riabchikova, G. G., 157(114), *167*
Riecke, F. F., 93(16), *117*
Riesz, P., 70(109), *86*
Riley, J. F., 34(32), *43*, 51(37), 52(37), *83*, 93(18), 94(18), *117*

AUTHOR INDEX

Risse, O., 6, 8, *17*, *18*
Rocklin, S. R., 200(29), *207*
Roebke, W., 54(41), *83*
Roesh, W. C., 210(1), *224*
Roseveare, W. E., 7, *18*
Ross, A., 22(6), *42*
Rotblat, J., 155, *167*
Roth, E., 2(14), *16*
Rothschild, W. G., 64(87), 70(87), 71(87), *85*
Rothwell, A. E., 157(111), *167*, 224(20), *225*
Russell, J. C., 42(82), *45*, 127, 128(16), *162*

S

Saeland, E., 198(21), *207*
Saito, E., 109(80), *120*
Saldick, J., 199(28), 202, *207*
Salzman, A. J., 113(96), *121*, 128(24), *163*
Samuel, A. H., 13, *18*, 35, *44*, 46(1), *81*, 170, *188*
Samuni, A., 109(84, 85), *120*
Sanders, J., 22(3), *42*
Sangster, D. F., 105, *120*
Sangster, M., 14, *19*
Santar, I., 29, 31, 32(15a, 15c), *42*, *43*
Sarrah, D., 102(68), *120*
Saunders, R. D., 201(39), *208*
Sawai, T., 137, 140(61), *165*
Schenck, G. O., 33(31), *43*
Scheraga, H. A., 145(78), *165*
Scheuer, O., 4, *17*
Schmidt, K. H., 35(40), 40(40, 55, 60), *44*, *45*, 48, 54(54, 58), 56(58), 64(58), *83*, *84*, 197(19), 202, *207*, *208*
Scholes, G., 3(28), 14, *16*, *19*, 46(8), 70(*106*), 73(8), *82*, *86*, 100(65), 106(65, 74), 107 (65), *119*, *120*, 128(26), 139(68), 155(26), *163*, *165*, 177(21), *189*
Schuler, R. H., 27(8), *42*, 155(97), *166*, 198, 199(27), 200, *207*
Schwarz, H. A., 2(18), 13, 14, *16*, *19*, 40(64, 74), *45*, 46(10), 58, 59, 70(111), 71 (111), 75(134), 76(138), 78(134), 79(134), *82*, *87*, *88*, 99(59), 110(87), 112, 113(92, 96, 99), *119*, *121*, 128(20, 24, 26, 33),

129(40), 134(33), 136(58), 139(40, 66), 140 (58, 66), 150(88), 155(26), *162*, *163*, *164*, *165*, *166*, 170, 174, 177, 178(5), 179, 182, 185(9), *188*, *189*
Seddon, W. A., 15(96), *19*, 46(17), 56(68), *82*, *84*, 97(39, 40), *118*
Sehested, K., 98(43), 110(86), *118*, *120*, 202 (42), *208*
Senior, W. A., 33(23), *43*
Shaede, E. A., 34, *44*
Shaw, P., 100(65), 106(65), 107(65), *119*
Shiga, T., 93(12), *117*
Shubin, V. N., 2, *16*
Sibirskaia, G. K., 157(114), *167*
Sicilio, F., 93(13, 14), *117*
Sigli, P., 155, *166*, 198(24), *207*
Silini, G., 3(32), *16*
Sills, S. A., 40(76), *45*, 64(83), 70(83), 71 (83), *85*, 99(63), 102(63), *119*
Simić, M., 70(106), *86*, 198(26), *207*
Smith, D. R., 14, *19*, 46(11), 60(11), *82*
Smith, H. P., 7(59), *18*
Smithies, D. H., 64(88), 72(88), *85*, 128(39), *163*
Snoek, C., 22(3), *42*
Sokolov, U., 33(26, 27), *43*
Spinks, J. W. T., 2(17), *16*, 109(81), *120*, 201 (41), *208*
Spitsyn, V. I., 157(113), *167*
Spurný, Z., 218(9), *225*
Stein, G., 3(30), 13, 15(103), *16*, *18*, *19*, 20, 29, 33(26, 27), *42*, *43*, 46(4), 64(86, 92, 93, 95), 65(86, 93, 96), 70(86, 92, 95, 100, 101, 105, 107), 71(86, 100), 72(123), *82*, *85*, *86*, *87*, 113(98), *121*, 145, *165*, 202, *208*
Stott, D. A., 70(110), *87*, 94(28), 99(28), 107 (28), *118*
Sutton, H. C., 13, *18*, 94(22, 25), 99(25), 113(22), *117*, 128(30), 140(30), 155, *163*, *167*
Sutton, J., 73(130), 74, 75(*132*, *135*), 76 (130), *87*, 88, 128(34), *163*, 198(22), *207*
Swallow, A. J., 2(15), 3(26), 15(26), *16*, 46 (15), 54(50), 55(50), 56(75), 64(50, 75, 85), 71(85), *82*, *84*, *85*, 94(24), 109(79), *117*, *120*, 130(48), 131(48), *164*, 201(39), *208*
Sweet, J. P., 40(65), *45*, 64(84), 70(84), 71 (84), *85*
Swinnerton, J. W., 143(73), *165*

Sworski, T. J., 13, 19, 34(33), *43*, 78(144), 79(145), *88*, 94(29,30), 111(90), 113(29, 30, 93), 115(90), *118, 120, 121*, 128(18, 19, 23), 129(18, 19), 154(96), *162, 163, 166*

T

Taimuty, S. I., 219(10), *225*
Tanaka, I., 33(29), *43*
Taub, I. A., 40(59), *45*, 54(55), 55(55), 67, 80, *84*, 96(34), 97, 100(34), *118*
Tazuke, S., 14(88), *19*, 46(11), 60(11), *82*
Teply, J., 15(97), *19*
Thomas, J. K., 15(99), *19*, 40(53, 54, 65, 67, 71, 75, 77), *44, 45*, 46(20), 54(45, 46, 47, 53), 55(46, 47, 53), 56(45, 46, 70), 57(53), 64(53, 82, 84), 70(82, 84, 112, 113), 71(46, 82, 84, 112), 81, *82, 83, 84, 85, 87, 88*, 93(17), 94(23), 99(17, 55, 58, 60), 100 (58, 60), 107(60), 112(58), 113(17), *117, 119*, 131(52), 155, 156, *164, 167*, 181, *189*, 195(8), 197(8), *206*
Timnick, A., 153(91b), *166*
Topp, M. R., 196(14), *207*
Towle, L. H., 219(10), *225*
Trumbore, C. N., 106(73a), *120*, 128(27), *163*

V

Van Cleave, A. B., 201(41), *208*
Vereshchinskii, I. V., 2(16), *16*
Vermeil, C., 33(28, 30), *43*
Verrall, R. E., 33(23), *43*
Vitushkin, N. I., 201(40), *208*
Vladimirova, M. V., 153, *166*

Vodar, B., 29(16), *43*
Voevodskii, V. V., 111(89), 115(89), *121*

W

Walker, D. C., 34, *44*, 46(19, 23), 81(148), *82, 88*
Wall, L. A., 93(13, 14), *117*
Wander, R., 97(37), 100(37), *118*
Watanabe, K., 29(17), *43*
Watt, W. S., 60, *85*
Weeks, J. L., 94(*26*), 99(26, 53), 102(53), *118, 119*, 131(53), *164*
Weiss, J., 12, 14, *18, 19*, 27(9), *42*, 46(6), 60 (6), 61, *82*, 102(67), *120*, 139(68), *165*
Wigg, E. E., 129(43), *164*
Wigger, A., 97(36), 100(36), *118*
Wilkinson, F., 56(68), *84, 97, 118*
Wilkinson, J., 200(32), *208*
Willis, C., 113(97), *121*, 128(29), 157, *163, 167*, 224(20), *225*
Willson, R. L., 95(32), 100(65), 106(65, 74), 107(65), *118, 119, 120*, 130(45), *164*
Wingate, C., 29(18), *43*
Wiseall, B., 106(73b), *120*
Wolff, R. K., 36(48), *44*, 51(36), *83*
Woods, R. J., 2(17), *16*
Woolsey, G. B., 126(8), *162*

Z

Zagorski, Z. P., 98(43), *118*
Zatulovskii, V. I., 201(40), *208*
Zeltmann, A. H., 70(104), *86*

Subject Index

A

Absorbed dose
  chemical change as measure, 211–215
  definition, 210
Absorption coefficient for water
  atomic, 24–25
  Compton, 24–25
  linear, 25
  pair production, 24
  photoelectric, 24–25
  total, 24
Absorption of radiation in water
  electrons, 26–27
  heavy charged particles, 27–28
  γ rays, 23–25
  neutrons, 28
  X rays, 23–25
Absorption spectrum
  hydrated electron, 50–51
  hydrogen atom, 70–73
  hydroperoxyl radical, 109–110
  hydroxyl radical, 93–94
Abstraction of hydrogen
  reactions of H, 70
  of OH, 95
Accelerators
  Cockroft–Walton, 198, 199, 200
  cyclotron, 198, 200
  linear, 194, 197, 198
  pulse generator, 195
  Van de Graaf, 194, 195

Acetone, 56, 64, 72, 74, 114
Acrylamide
  $+e_{aq}^{-}$, 60
  $+Fe^{3+}$, primary yield determination, 146
  $+OH$, 114
  yield of primary products in H_2O and D_2O, 146
Activation energy for reactions of hydrated electron, 159
"Activated water," Fricke's concept, 5, 8
Active species in water
  absolute yields, 31–32
  spatial distribution, 36–38
Alpha rays
  absorption in water, 27–28
  early experiments with, 3–5
  from nuclear reactions, 198
  polonium, 197
  radon, 9
Atomic hydrogen, 63–76
  absorption spectrum, 70–73
  $+Ce^{4+}$, 12
  $+(CH_3)_2CDOH$, 141
  $+(CH_3)_2CHOH$, 137
  $+ClCH_2COOH$, 60
  comparison of rate constants for reactions of hydrated electron, 64
  conversion to hydrated electron, 68–70
  $+Cu^{2+}$, 70, 71
  diffusion constant, 179, 182
  $+e_{aq}^{-}$, 40, 76, 178, 182
  $+F^{-}$, 69

+HCOO$^-$, 130
+H$_2$O$_2$, 11, 63, 92, 179, 182
+H$_3$O$^+$, 64, 70
+HPO$_3^{2-}$, 136
hydrated electron conversion to, 36, 40, 52, 65, 73
isotopic effects in relative rate constants of, 149
+O$_2$, 12, 63, 92, 108, 130, 161
+OH, 11, 12, 40, 178, 182
OH radical conversion to, 11, 40, 68
+OH$^-$, 68, 69, 70
origin, 33, 36, 73–76
production by high-frequency discharge, 202
rate constants with inorganic solutes, 71
with organic solutes, 72
ratios of radiation-chemical yields in H$_2$O and D$_2$O, 148
recombination, 11, 40, 76, 178, 182
role in radiolysis, 63–65
theoretical predictions on dependence of primary yield on scavenger reactivity, 185–186
yield in heavy water, 149
in neutral solutions, 138, 139, 140, 141
Aqueous chemical dosimeters, 211–225, *see also* Dosimeters

B

Benzoic acid, 97, 105–106
Beta rays, *see also* Electron
early experiments with, 3–5
G(Fe^{3+}) for, 218
internal sources of, 200–201
in nuclear reactor radiations, 201
of tritium and primary radical yields, 154
Bicarbonate ion, 161
Bimolecular reaction
pseudo-first-order, 66–68
second-order, 95–97
Blob, 37–38
Boron, reaction with neutrons, 151, 155, 198
Bromide ion, 113, 146

C

Carbon dioxide
from benzoic acid, 105

+e$_{aq}^-$, 131, 161
from HCOONa + O$_2$, 127–128, 134
Carbon monoxide, 54, 137
Carboxyl radical, 130–132
Ceric sulfate
α-radiolysis, 153
dosimeter, 213, 219–220
+H$_2$O$_2$, 109
LET effect on G(Ce^{3+}), 153
Chemical kinetics
competition method, 103–107
contribution of radiation chemistry, 2
diffusion-kinetic model, 171–189
pseudo-first-order reaction, 66–68
second-order reaction, 95–97
steady-state method, 132
Chemical stage of radiolysis, 39
Chloride ion, 113
Chloroacetic acid
+e$_{aq}^-$, 60, 68
+H, 60
stable products yield, dependence on concentration, 61
on pH, 62
Cobalt-60 radiation units, 191–193
Competition kinetics, 103–107
Compton effect, 24–25
Conductivity during and after irradiation, 48
Copper ion
addition to Fricke dosimeter, 213
in D$_2$O, 146
+e$_{aq}^-$, 53, 149
+H, 70
+HO$_2$, 108
Cyclohexadienyl radical, 97
Cyclotron, 198, 200

D

Deuteron
absorption in water, 27–28
transfer, 149–150
Diffusion constant,
atomic hydrogen, 179, 182
atomic oxygen, 182
H$_3$O$^+$, 179, 182
hydrated electron, 48, 179, 182
hydrogen peroxide, 179, 182
molecular hydrogen, 179
OH, 179, 182
OH$^-$, 179, 182

SUBJECT INDEX 237

Diffusion kinetic model, 171–189
  basic assumptions, 171–172
  choice of parameters, 175–177
  Kuppermann's many radical, 181–188
  one-radical, 173
  one radical and one-solute, 173–175
  Schwartz many radical, 178–181
  theoretical predictions and experimental verifications, 177–178
  two radicals and one solute, 175
D_2O, see Heavy water
Dose rate effect, 155–157
  on $G(Fe^{3+})$, 156
  on molecular hydrogen yields, 157
Dosimeters
  adiabatic calorimeter, 224
  benzene, 213
  ceric sulfate, 213, 219–220
  differential calorimeter, 224
  dyes, 213
  ferrous-cupric, 213
  ferrous sulfate (Fricke dosimeter), 6, 213, 216–219
  hydrocarbons, releasing HCl, 213
    optically active, 214
  oxalic acid, 213, 220–222
  potassium ferrocyanide, 214, 224
  potassium thiocyanate, 214, 223–224
  sodium formate, 213
  water (hydrated electron), 214, 222–223

E

e_{aq}^-, see Hydrated electron
Electron, see also Hydrated electron, Beta rays
  accelerator, 194–195
  energy loss, 26–27
  fast, 26
  fate in water, 35–36
  hydration, 39, 50
  slow, 26
  subexcitation, 26, 35
  thermal, 26
  thermalization, 39
Electron transfer, 53, 94
Ethanol
  in D_2O, 80, 146

$G(H_2O_2)$ in air saturated solution, 136
  scavenger for H, 73–74
  for OH, 67, 114
α-Ethanol radical
  absorption spectrum, 96
  molar extinction coefficient, 96
  recombination, 95–97
Excitation potentials of water, 27, 29–30

F

Fenton's reagent, 92
Ferric ion, see also Ferrous ion oxidation
  molar extinction coefficient, 216
  reduction in acrylamide solution, 146
Ferrous ion oxidation
  dose rate effect, 155–156
  Fricke dosimeter, 6, 213, 216–219
  $G(Fe^{3+})$ in D_2O, 147
    in H_2O, 136, 151, 156
  with H_2O_2, 92
  LET effect, 151–153
  with OH, 12, 94, 136
  reaction scheme, 136
Ferricyanide ion, 60, 113, 141, 224
Ferrocyanide ion, 112, 113, 214, 224
Fluoride ion, reaction with H atom, 69
Formic acid + oxygen, aqueous solutions
  effect of LET, 153–154
  mechanism of radiolysis 130–135
  primary yields in D_2O, 145, 148–149
    in H_2O, 135–136, 142–143
  stable radiolytic products, 128, 133–134
Free-radicals, primary, see also individual radicals
  formation, 38–42
  model of radiolysis, historical, 9–14
  spatial distribution, 36–38
  time scale of various processes, 38–42
Fricke dosimeter, 6, 213, 216–219
Gamma rays
  absorption in water, 23–25
  in nuclear reactor radiations, 201–202
  sources of, 191–193

G

G-value, see Radiation-chemical yield

H

H_2^+, 70
Heavy charged particles
  absorption in water, 27–28
  from accelerators, 198–200
  from nuclear reactions, 198
Heavy water
  decomposition yield, 148–149
  isotopic effect in relative rate constants, 149–150
  primary yields, 144–150, 154
  rate constant with e_{aq}^-, 53, 150
  ratios of radiation chemical yields in H_2O and D_2O, 148
H_3O^+
  diffusion constant, 179, 182
  $+e^-$, 11, 36
  $+e_{aq}^-$, 36, 58, 60, 65, 68, 71, 73, 140, 178, 181, 182
  in ferrous ion oxidation, 70, 161
  formation, 10, 34, 36, 39, 91, 101
  $+H$, 70
  in heavy water, 149
  $+OH^-$, 75
  radius of spur, 179, 182
  yield, 35, 179
Hydrated electron
  absorption spectrum, 50–51
  $+Ag^+$, 60
  apparatus for photogeneration of, 202–203
  cavity radius, 50
  charge density radius, 49
  $+CH_2{:}CHCONH_2$, 60
  $+ClCH_2COOH$, 60, 68
  $+CO_2$, 131, 161
  conversion to OH, 93
  $+C(NO_2)_4$, 51
  diffusion constant, 48, 179, 182
  dimer, 81
  discovery, 14
  dosimeter, 214, 222–223
  energy levels, 50
  evidence for negative charge of, 57–60
  $+Fe(CN)_6^{3-}$, 60
  formation, 35–36, 50, 69
  $+H$, 40, 76
  half-life of, 47, 52–53
  $+HCOO^-$, 130
  $+H_2O$, 52
  $+H_2O_2$, 58, 63, 93
  $+H_3O^+$, 58, 60, 65, 73, 140
  hydration energy, 49
  ionic strength effect on rate constants, 57–60, 77
  molar extinction coefficient, 51
  $+NO_2^-$, 58
  $+NO_3^-$, 76–79, 161
  $+N_2O$, 60, 93
  $+O_2$, 58, 63, 108, 130
  $+OH$, 40, 115
  properties of, 47–57
  radius of spur, 134, 179, 182
  rate constants with inorganic solutes, 54–55
    with organic solutes, 56
  reactions with primary free-radicals, 40
  reactivity, 51–56
  recombination of, 76, 80–81, 178, 182
  redox potential, 48–49
  relaxation of ionic atmosphere, 47
  schematic picture of polarization, 50
  simplest negative ion, 47–50
  theoretical predictions on dependence of yield on scavenger reactivity, 185–186
  yield in heavy water, 148, 149
    in neutral and alkaline solution, 140
Hydrogen, *see* Atomic hydrogen, Molecular hydrogen
Hydrogen peroxide
  dependence on dose rate, 156
  $+e_{aq}^-$, 58, 63, 93
  $+Fe^{2+}$, 92
  formation, 4, 8, 9, 11, 110–113, 115–116, 130–132, 136
  $+H$, 11, 92
  $+HPO_3^-$, 137
  $+h\nu$, 92
  $+OH$, 92, 108, 152
  $+Ti^{3+}$, 92
  yield as check of water purity, 204
  in oxygenated solutions
    effect of formate ion concentration, 133–134
    formation, 110, 130–132, 136
    in solution of $H_2 + O_2$, 62–63
    yields of, 134, 139–140, 204
  primary
    dependence on LET, 152–154

SUBJECT INDEX

on pH, 142
on reactivity, 113–115
on temperature, 158
formation, 111–113, 115–116
theoretical predictions on dependence of yield on scavenger reactivity, 185–186
yield of, 135, 139–140
in heavy water, 148–149
Hydroperoxyl radical, 108–111
absorption spectrum, 109
$+Ce^{4+}$, 109
$+Cu^{2+}$, 108
ESR identification of, 109
formation of, 12, 92, 108, 109, 152
LET effect on yield of, 152–154
nature at various pH, 110
$+OH$, 98
pK value of, 110
as primary species, 108, 127, 152
properties, 108–109
rate constants, 110
recombination reaction, 110, 131
redox potential, 108–109
as secondary radical, 108
Hydroxyl radical
absorption spectrum, 93–94
abstraction of H, 95
addition reactions, 97
adduct, 97–98
$+(CH_3)_2CHOH$, 137
$+C_6H_6$, 97
$+CNS^-$, 107
conversion to H, 68
to O^- form, 102–103
differences in behavior of O^- and OH form, 102–103
$+e^-_{aq}$, 40, 115, 178, 182,
electron transfer reactions of, 94
$+Fe^{2+}$, 12, 94, 136
$+Fe(CN)_6^{4-}$, 112
formate ion reaction with O^- form of, 140
formation of, 4, 8, 10, 33–34, 39, 92–93
$+H_2$, 11, 63, 92
$+HCOO^-$, 130
$+HO_2$, 98
$+H_2O_2$, 92, 108, 152
$H_2O^+_{aq}$ as acid form, of, 101–102
$+HPO_3^-$, 136

$+HPO_3^{2-}$, 136
isotopic effect in relative rate constants of, 149
kinetic salt effect, 101
molar extinction coefficient, 93
of O^- form of, 102
$+O_2$, 98
oxygen reaction with O^- form, 98, 142
$+OH^-$, 102, 140
rate constants with inorganic solutes, 99
with organic solutes, 100
reactivity, 94–101
reactions with primary free-radicals, 40
recombination reaction, 11, 40, 112–113, 143, 179, 182
redox potential, 91
theoretical predictions on dependence of yield on scavenger reactivity, 185–186
yield dependence on LET, 152–155
on pH, 142
on temperature, 157–158
yield in heavy water, 148, 149
in neutral water, 135, 138, 140

I

Inorganic free-radicals, secondary
Br, 94
Cl, 94
CNS, 107
H_2O_3, 98
I, 94
Interconversion of H and e^-_{aq}, 52, 65, 68–70
Iodide ion, 113
Ionic strength effect
e^-_{aq} absolute rate constants, 57–58
e^-_{aq} relative rate constants, 58–60
OH relative rate constants, 101
Ionizing radiation, *see also* Electron, Gamma rays, Heavy charged particles, Neutrons, X rays
interaction with water, 23–28
sources of, 191–206
Inorganic scavengers
rate constant with e^-_{aq}, 54–55
with H, 71
with OH, 99
Ionization potential of water, 29, 34

Irradiation cell
  pulsed electron beam experiments, 195–196
  positively charged particle experiments, 199–200
Linear energy transfer, *see* LET

L

LET
  definition, 150
  effect on primary radical and molecular yields, 153
  on water decomposition, 150–155
  $G(Fe^{3+})$ dependence on, 151
  theoretical curve for water decomposition yield dependence on, 184
Lithium, reaction with neutrons, 151, 198, 215

M

Material balance for water decomposition, equation, 126–127, 132, 154
Molar extinction coefficient
  Ce^{4+}, 220
  $(CNS)_2^-$, 223
  copper–benzidine reagent for oxalic acid, 221
  α-ethanol radical, 96
  Fe^{3+}, 216
  $Fe(CN)_6^{3-}$, 224
  H atom, 72
  HO_2, 109
  H_2O_3, 98
  hydrated electron, 51
  O^-, 102
  O_2^-, 109
  O_3^-, 98
  OH, 93
Molecular hydrogen
  dependence on dose rate, 156
  formation of, 4, 11, 33, 61–62, 70, 141
  reactions with primary free-radicals, 40
  yield as check of water purity, 204
  primary
    dependence on dose rate, 156
    on LET, 153–154
    on pH, 142
    on reactivity, 76–80
    on temperature, 158
  in heavy water, 148–149
  theoretical predictions on dependence of yield on scavenger reactivity, 185–186
  yield of, 76, 135, 140, 142, 179

N

Neutrons
  absorption in water, 28
  nuclear reactions, 198
  in nuclear reactor radiations, 201–202
Nitrate ion
  concentration effect on atomic hydrogen yield, 73–74
    on molecular hydrogen yield, 76–80
  concentrated solutions of, 129
  effect on yield of primary H_2O_2, 114–115
Nitrite ion, 58, 113, 143
Nitrous oxide, 60, 93, 127–128, 141
Nuclear reactor
  dosimetry in, 221–222, 224
  source of radiation, 201–202

O

O^- ion-radical, *see* Hydroxyl radical
OH radical, *see* Hydroxyl radical
Optical approximation
  rule of, 30–31
  spectrum of water, 31
Organic scavengers
  rate constants with e_{aq}^-, 56
  with H atom, 72
  with OH, 100
Organic free-radicals
  carboxyl, 130–132
  cyclohexadienyl, 97
  α-ethanol, 95–97
Oxalic acid
  decomposition, 7
  dosimeter, 220–222
  primary yields in H_2O and D_2O from, 146
  reaction with e_{aq}^-, 56

SUBJECT INDEX

Oxygen
　+CO, 137
　+e_{aq}^-, 58, 108, 130
　+Fe^{2+}, 136, 147, 151–153, 155–156
　formation in ferricyanide + ferrocyanide system, 141
　+formic acid, 128, 130–132, 142, 145, 148–149
　+H, 108, 130, 161
　+O^-, 98, 142
　reaction with organic radicals, 95
Ozonide ion, 98

P

pD effect, 148–149
pH dependence
　hydroxyl radical form, 101–103
　primary radical and molecular yields, 140–144
　stable product yields of $ClCH_2COOH$ aqueous solutions, 61–62
Phosphite, in primary yield determination, 136–137
Physical stage of radiolysis, 38–39
Physicochemical stage of radiolysis, 39
PNDA (p-nitrosodimethylaniline), 106, 138
Pressure, effect on yields, 159–161
1-Propanol, 114
2-Propanol
　reactions with atomic hydrogen, 73–74, 137
　with OH, 137
2-D-2-Propanol, 141
Pseudo-first-order reaction, example, 66–68
Pulsed electron beam
　experimental arrangements, 194–197
　historical, 14–15

R

Rad, definition, 210
Radiation chemical yield
　corrections for increased reactivity, 133–135
　definition, 13
　dependence on dose rate, 155–157
　stable products in $HCOO^- + O_2$ system, 133, 135, 142–143
primary species
　calculation from experimental data, 130–132
　definition, 41–42, 123–125
　dependence on LET, 150–155
　　on pH, 140–144
　　on pressure, 159–161
　　on scavenger reactivity, 127–129, 133–135
　　on temperature, 157–159
　　on various factors, 123–168
　determined from $(CH_3)_2CDOH + N_2O$ system, 141
　from $(CH_3)_2CHOH + NO_3^-$ system, 137
　from $CO + O_2$ system, 137
　from $Fe^{2+} + O_2$ system, 135–136, 141
　from $HPO_3^{2-} + NO_3^-$ system, 136–137
　from $K_4Fe(CN)_6 + K_3Fe(CN)_6$ system, 141
　from PNDA, 138
　heavy water, 144–149
　neutral pH, 140
　theoretical predictions on dependence on scavenger reactivity, 185–186
Radiation chemistry of water
　early period, 3–9
　historical, 1–20
　present status, 1–2
Radiation sources, 191–203
　beta rays, internal, 200–201
　^{60}Co gamma rays, 191–193
　electron beam
　　continuous, 202
　　pulsed, 194–197
　intermittent ^{60}Co radiations, 201
　nuclear reactor, 201–202
　positively charged particles, 197–200
　X rays, 201
Rate constants
　comparison for e_{aq}^- and H reactions, 64
　e_{aq}^- with inorganic solutes, 54–55
　　with organic solutes, 56
　H atom with inorganic solutes, 71
　　with organic solutes, 72
　HO_2 and O_2^-, 110

ionic strength effect on absolute, 57–58, 77
   on relative, 58–60, 101
OH and O⁻ with inorganic solutes, 99
   with organic solutes, 100
reactions of primary free-radicals, 40
Scavenger reactivity effect, 127–129, 133–135
Secondary electrons, 35–36
Second-order reaction, example of, 95–97
Short tracks, 37–38
Spur
   distribution of absorbed energy as function of size, 180
   formation of, 37–38
   initial number of radicals in, 175–177, 180, 182
   ionic strength, initial, in, 180
   radius, 134, 175–177, 179, 182
Steady-state kinetics, 132
Subexcitation electrons, 26, 35, *see also* Electron
Superexcitation of water, 30, *see also* Water, excited
Syringe technique, 205–206

T

Temperature
   effect on primary radical and molecular yields, 158
      on reaction rate constants, 158–159
Tetranitromethane, 51
Thallous ion, 113
Thiocyanate, 103, 106, 107, 214, 223–224
Time scale of events in irradiated water, 41
Titanous ion, 92
Uranium, reaction with neutrons, 198

W

Water
   excited
      absolute yield, 31–32
      contribution to formation of primary products, 29
      decomposition, 11, 33, 39, 41
      formation, 29–30, 35
      importance for primary molecular hydrogen formation, 33, 78–79
         for primary H_2O_2 formation, 115–116
      triplet state, 29, 34
      superexcitation, 30
   ionized
      absolute yield of, 31–32
      formation, 4, 10–11, 30
      mass-spectrometric data, 30
   purification
      check of, 204
      degassing, 204–205
      distillation, 203–204
   vapor, irradiations, 33–34

X

X rays
   absorption in water, 23–25
   "photochemical" reactions induced by, 5–9
   sources, 201

Y

Yield, *see* Radiation chemical yield

Physical Chemistry

A Series of Monographs

Ernest M. Loebl, Editor

Department of Chemistry, Polytechnic Institute of Brooklyn, Brooklyn, New York

1. W. JOST: Diffusion in Solids, Liquids, Gases, 1952
2. S. MIZUSHIMA: Structure of Molecules and Internal Rotation, 1954
3. H. H. G. JELLINEK: Degradation of Vinyl Polymers, 1955
4. M. E. L. MCBAIN and E. HUTCHINSON: Solubilization and Related Phenomena, 1955
5. C. H. BAMFORD, A. ELLIOTT, and W. E. HANBY: Synthetic Polypeptides, 1956
6. GEORGE J. JANZ: Thermodynamic Properties of Organic Compounds — Estimation Methods, Principles and Practice, revised edition, 1967
7. G. K. T. CONN and D. G. AVERY: Infrared Methods, 1960
8. C. B. MONK: Electrolytic Dissociation, 1961
9. P. LEIGHTON: Photochemistry of Air Pollution, 1961
10. P. J. HOLMES: Electrochemistry of Semiconductors, 1962
11. H. FUJITA: The Mathematical Theory of Sedimentation Analysis, 1962
12. K. SHINODA, T. NAKAGAWA, B. TAMAMUSHI, and T. ISEMURA: Colloidal Surfactants, 1963
13. J. E. WOLLRAB: Rotational Spectra and Molecular Structure, 1967
14. A. NELSON WRIGHT and C. A. WINKLER: Active Nitrogen, 1968
15. R. B. ANDERSON: Experimental Methods in Catalytic Research, 1968
16. MILTON KERKER: The Scattering of Light and Other Electromagnetic Radiation, 1969
17. OLEG V. KRYLOV: Catalysis by Nonmetals — Rules for Catalyst Selection, 1970
18. ALFRED CLARK: The Theory of Adsorption and Catalysis, 1970

Physical Chemistry

A Series of Monographs

19 ARNOLD REISMAN: Phase Equilibria: Basic Principles, Applications, Experimental Techniques, 1970
20 J. J. BIKERMAN: Physical Surfaces, 1970
21 R. T. SANDERSON: Chemical Bonds and Bond Energy, 1970
22 S. PETRUCCI, ED.: Ionic Interactions: From Dilute Solutions to Fused Salts (In Two Volumes), 1971
23 A. B. F. DUNCAN: Rydberg Series in Atoms and Molecules, 1971
24 J. R. ANDERSON: Chemisorption and Reactions on Metallic Films, 1971
25 E. A. MOELWYN-HUGHES: Chemical Statics and Kinetics of Solution, 1971
26 IVAN DRAGANIĆ AND ZORICA DRAGANIĆ: The Radiation Chemistry of Water, 1971

In Preparation

M. B. HUGLIN: Light Scattering from Polymer Solutions

ASHEVILLE-BUNCOMBE TECHNICAL COLLEGE

3 3312 00010 5825